ASTRONOMY

ACTIVITY AND LABORATORY MANUAL

ALAN W. HIRSHFELD

UNIVERSITY OF MASSACHUSETTS DARTMOUTH

JONES AND BARTLETT PUBLISHERS

Sudbury, Massachusetts

BOSTON   TORONTO   LONDON   SINGAPORE

World Headquarters

Jones and Bartlett Publishers
40 Tall Pine Drive
Sudbury, MA 01776
978-443-5000
info@jbpub.com
www.jbpub.com

Jones and Bartlett Publishers
Canada
6339 Ormindale Way
Mississauga, Ontario L5V 1J2
Canada

Jones and Bartlett Publishers
International
Barb House, Barb Mews
London W6 7PA
United Kingdom

Jones and Bartlett's books and products are available through most bookstores and online booksellers. To contact Jones and Bartlett Publishers directly, call 800-832-0034, fax 978-443-8000, or visit our website www.jbpub.com.

ISBN-13: 978-0-7637-6019-9

Production Credits

Publisher: Cathleen Sether
Managing Editor: Dean W. DeChambeau
Acquisitions Editor: Shoshanna Goldberg
Associate Editor: Molly Steinbach
Editorial Assistant: Caroline Perry
Production Director: Amy Rose
Associate Production Editor: Melissa Elmore
Senior Marketing Manager: Andrea DeFronzo
V.P. of Manufacturing and Inventory Control: Therese Connell
Composition: Spoke & Wheel/Jason Miranda
Cover Design: Brian Moore
Associate Photo Researcher and Photographer: Christine McKeen
Cover Image: © Courtesy of NASA
Chapter Opener: Courtesy of NASA and H. Richer (University of British Columbia)
Printing and Binding: Courier Stoughton
Cover Printing: Courier Stoughton

6048

Printed in the United States of America

12 11 10 09 08   10 9 8 7 6 5 4 3 2 1

Contents

Introduction

Welcome to the universe. According to publishers' figures, every year some 150,000 under-graduates across the nation take an introductory astronomy course. Humankind has always been curious about "what's out there," how it all began, and how we, humanity, fit into this mind-boggling cosmic scheme. And astronomers have always responded to this curiosity, delivering a steady stream of discoveries about the universe, its objects, and its processes. When I was a college student, back in the early 1970s, our knowledge of the universe seemed frustratingly hazy. Telescopes weren't quite big enough, computers weren't nearly fast enough, and our knowledge of physics wasn't deep enough to answer many long-standing questions about planets, stars, galaxies, and the universe. Times have changed—rapidly and drastically. The remarkable acceleration of technology has propelled humanity to a level of understanding that dazzles even the most solemn astronomer. And I have the privilege of introducing this story to others.

I have taught introductory astronomy to thousands of science and liberal arts students since I joined the Physics Department at the University of Massachusetts at Dartmouth in 1978. Until recently, I had presented the subject matter in the "standard" format: a lecture with appropriate visual aids. However, it became increasingly clear to me that this passive instruc-tional approach failed to engage many of my students, a suspicion confirmed by their inat-tentiveness and subpar test performances. At the same time, I became involved in UMass Dartmouth's innovative freshman engineering curriculum—the IMPULSE program—in which students learn through guided, team-based problem-solving and laboratory activities, rather than from lectures. This active approach has proven very successful, even with a sub-ject as difficult as physics, increasing student attentiveness, retention, and standardized test scores. The IMPULSE experience, plus my background as a writer of popular-level articles and books on science history, provided the dual inspiration for me to completely overhaul my introductory astronomy course. This activities manual is one result.

My new attitude to teaching came, ironically, by temporarily setting aside my identity as a teacher. Instead, I considered my work in the classroom from a writer's perspective, and immediately realized what was missing from the standard introductory college astronomy course: the essential storyline that tells how astronomers came to know as much as they do about the universe. When the human element is removed from the narrative, the story itself withers. Many students find it hard to engage with a sterile presentation of scien-tific principles and knowledge, no matter how logically organized, visually compelling, or energetically delivered. The course becomes a catalog of seemingly isolated facts, figures, and taxonomy, divorced from the thrilling story of scientific advancement. Many of today's textbooks make, at best, a token effort to reveal the historical arc of astronomical discovery. By contrast, I see an integrated historical narrative as the very oxygen that breathes life into the study of astronomy.

Through this manual's series of 20 modestly mathematical, paper-and-pencil classroom activities, you will recapitulate the epic advancement of astronomical thought, from the rudimentary observations of prehistoric skywatchers to the development of modern astrophysics in the 20th century. You will repeat essential elements of the work of history's most noted astronomers. And, although groundbreaking in its time, their work is accessible to today's college student—you don't have to be an Isaac Newton to retrace some of his thinking. By experiencing for yourself a condensed version of the rise of knowledge about our universe, I hope you will better appreciate the fruits of modern research. Every new discovery, no matter how technical or abstruse, has its roots in humanity's age-old quest to comprehend the cosmos.

I developed and implemented these astronomy activities at UMass Dartmouth during the Fall 2006 semester in three introductory astronomy classes totaling 240 non-science students. Since then, I have refined the activities based on student input. My students work in a regular lecture hall in teams of two to four, with me and a teaching assistant on hand to answer questions. This manual can be used either as a stand-alone core of an activities-oriented astronomy course or as a supplement to an existing textbook.

The activities in this manual require no specialized equipment or individual materials beyond a pencil, a straightedge, and a common calculator. They are designed for use in a classroom of any size, from a small seminar room to a large lecture hall. The necessary mathematical background—basic elements of high school algebra, geometry, and trigonometry—is introduced on an as-needed basis for each activity. And if you count yourself among the "math phobic," don't worry; with few exceptions, my students were able to work through these activities successfully.

The World's First Skywatcher—YOU!

To prehistoric observers, skywatching must have been regarded as necessary for survival—for both rational and irrational reasons. In the face of what must have seemed an unpredictable and often hostile environment, early humans sought ways to organize, to understand and, through various superstition-based rites, to influence natural occurrences. The starting point for such a task would have been to observe phenomena around them in order to gradually assemble a "database" of information about natural patterns and cycles. For example, astronomical observation was so important to the ancient Mayans that they erected huge stone pyramids to serve as elevated observing platforms. While the earliest skywatchers certainly had both astrological and religious motivation, people eventually realized that knowledge of the sky also made local direction-finding, navigation, and calendar-keeping possible.

1. Imagine yourself living in prehistoric times, when the only astronomical instrument was the human eye. What kinds of astronomical phenomena would you, a prehistoric skywatcher, have seen in the heavens above? Consider both the daytime sky and the nighttime sky, and assume that the air above you is cloudless. Write down as many astronomical phenomena as you can think of, with a brief description of what you would see, on the worksheet. Be ready to discuss your answers in class.

2. Suppose you were to step outside tonight to observe the night sky. Given the considerable changes to your local environment over time, how might your view of the night sky differ from what one presumably would have seen during prehistoric times?

3. Where in the world might you travel to view the night sky as it appeared to prehistoric skywatchers? Explain your reasoning.

The World's First Skywatcher—YOU!

All work must be shown to receive credit.

1. _____

2. _____

_____

_____

_____

3. _____

_____

_____

_____

Shadowland

One of the first astronomical measuring instruments was the *gnomon*, which in its simplest form is a straight stick placed vertically into the ground. The gnomon is used to track both daily and longer-term movement of the Sun in the sky. Couple the gnomon with a graduated scale of the hours and you have a sundial. To understand the genesis of the gnomon, consider which general features of the Sun's movement ancient observers would have noticed:

- Each day, the Sun rises somewhere in the eastern half of the sky and sets somewhere in the western half.

- Shadows are longer during the early mornings and late afternoons when the Sun is low in the sky; they are shorter around midday when the Sun is high.

- The Sun climbs higher in the sky during summer than during winter.

- Summer days have more hours of daylight than winter days; that is, the Sun rises earlier and sets later during summer.

The gnomon was probably introduced in an effort to quantify these general observations, that is, to measure the Sun's movement in a consistent and reliable way. As a result, the Sun could serve both as a daily clock and as a way to track the passage of the seasons, providing a rational way for people to decide such essential matters as when to plant crops, when to store food, and when to prepare shelter. As you can see, early civilizations had very practical and even life-sustaining reasons to pursue the study of astronomy.

Here is how the gnomon was used. The gnomon was stuck vertically into the ground in a flat, open area where the Sun could shine on it all day long. As the day progressed, the observer placed markers—say, pebbles—to identify where the tip of the gnomon's shadow fell on the ground. By day's end, the pebbles would have traced out an arc. What was the shape of such an arc? The long shadows during early morning and late afternoon meant that the pebbles near the ends of the arc were relatively far from the gnomon. Pebbles placed around midday, when the Sun was high and cast shorter shadows, were closer to the gnomon. In fact, the shortest shadow—the point on the arc closest to the gnomon— indicated the time of midday, local noon. And, if you think about it, because the shortest shadow also indicated the local north–south direction, the gnomon also served as a primitive geographic compass. Not bad for a simple stick!

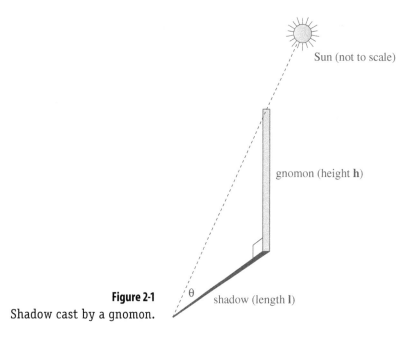

Sun (not to scale)

gnomon (height **h**)

Figure 2-1
Shadow cast by a gnomon.

θ   shadow (length **l**)

In **Figure 2-1**, notice how the gnomon and its shadow form a right triangle. The hypotenuse—the triangle's longest side—is the imaginary line segment connecting the gnomon's tip to the shadow's endpoint. The angle the hypotenuse makes with the shadow is represented here by the Greek letter theta **θ**, which is the Sun's altitude in the sky. (Altitude is expressed as an angle, starting at zero degrees at the horizon and reaching a maximum of 90 degrees straight overhead.)

1. Simple trigonometry tells us that the tangent, or tan, of an angle is computed by taking the *length of the side opposite the angle* and dividing that by the *length of the side adjacent to the angle*. Use this formula as the basis to develop your own formula relating the Sun's altitude **θ**, the gnomon's height **h**, and the shadow's length **l**.

2. Use your formula from Part 1 to compute the altitude of the Sun (in degrees) when a 1-foot-tall gnomon casts a shadow ½-foot long. You will need to use your calculator's "inverse tangent" key, sometimes labeled $\tan^{-1}$. Make sure your calculator is set to accept and display angles in units of degrees, not radians; the calculator display should indicate DEG, not RAD. Ask for assistance if you need it.

3. How long would the shadow cast by a 1-foot-tall gnomon be if the Sun was 30 degrees above the horizon?

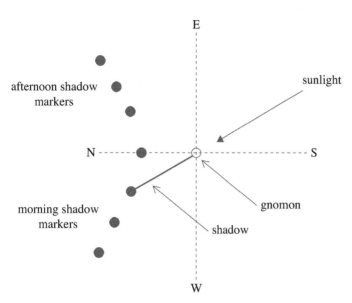

Figure 2-2
Gnomon shadow plot (top view).

During the course of the day, the Sun and the gnomon's shadow each move through a number of geographic quadrants. These quadrants are indicated in the accompanying shadow plot, **Figure 2-2**, which is for a mid-northern hemisphere location: *northeast, southeast, southwest,* and *northwest*. Use these quadrant designations to answer the questions below.

4. In which quadrant does the Sun appear in the morning?

5. In which quadrant does the gnomon's shadow appear in the morning?

6. In which quadrant does the Sun appear in the afternoon?

7. In which quadrant does the gnomon's shadow appear in the afternoon?

8. Write a general statement comparing the geographic direction of the Sun at any time relative to that of the gnomon's shadow.

9. Explain why the markers in the shadow plot are farther from the gnomon in the early morning and late afternoon compared to times near midday.

10. In which geographic direction does the gnomon's shadow point when the shadow is at its shortest during the day?

11. In which geographic direction is the Sun at this same time?

Shadowland

All work must be shown to receive credit.

1. _____

2.

3.

4. _____ quadrant

5. _____ quadrant

6. _____ quadrant

7. _____ quadrant

8. _____
_____
_____

9. _____

_____

_____

10. _____

11. _____

Shadowland—The Sequel

The shadow plot in Activity 2, Figure 2-2, depicts the shadow cast by a gnomon during a single day's transit of the Sun across the sky. Ancient observers constructed such daily plots over the course of many months and even years to record changes that occur with the passage of the seasons. Even though the gnomon has ancient roots, it can be used to outline the steps modern scientists take to organize and analyze observed data: first, array the data in a *table*; then, illustrate the data in a *graph*; and, finally, describe the overall pattern of the data mathematically as an *equation*. Scientists use such equations both to predict the outcome of related observations or experiments and to comprehend the underlying physics of natural phenomena.

1. Explain why a gnomon-based method of tracking the passage of the seasons might be more reliable than tracking the passage of the seasons by local weather conditions alone.

■ Constructing a Table of Data

A table is a straightforward way to organize a collection of data gathered from observations or experiments. **Table 3-1** on the worksheet shows shadow-length measurements that an ancient observer might have made. (The numbers are typical for a northern-hemisphere latitude of around 40 degrees.) The column labeled **Days** lists the number of days since "day zero," which is arbitrarily set to the first day of spring—around March 21. The column labeled **l** lists the shortest (noontime) shadow length for that given day. The shadow length is expressed here simply as a fraction or a multiple of the gnomon's height **h**, so it does not matter that the measurement rulers of ancient observers were different than those used today; that is, we can assume that the height of the gnomon **h** equals 1. The columns labeled **tan θ** and **θ** are, respectively, the tangent of the Sun's altitude and the Sun's altitude itself, in degrees. (For simplicity, the data has been adjusted to a hypothetical year that has 360 days, not 365.)

2. In Part 1 of Activity 2, you constructed a formula relating the Sun's altitude **θ**, the gnomon's height **h**, and the shadow's length **l**. Use this formula to compute both **tan θ** and **θ** for each of the **Days** listed in the table on the worksheet. Write your answers in Table 3-1.

3. Describe any patterns you see in the **Days** and **θ** data in Table 3-1.

■ Constructing a Graph of Data

A table is a useful way to organize data, but it is not the best way to reveal correlations, patterns, or periodicities in a mass of data. For this, a graph of the data is usually better. When the table is converted to a graph, numerical correlations that were hidden or only hinted at in the table often become plainly evident. Also, it may be possible to use the graph to *predict* data values other than those used to construct the graph.

4. Use the axes in **Figure 3-1** to plot a graph of Days vs. **θ** for the data in your table. There should be 1 data point corresponding to each table row, for a total of 9 data points.

5. On your graph, label the data points that correspond to the first day of summer (*summer solstice*) and the first day of winter (*winter solstice*).

6. Carefully draw a smooth curve connecting the data points. If you do this correctly, you should see a type of curve called a *sine wave*. The sine wave pattern would repeat year after year.

7. Use the curve you drew to predict (that is, read off) the altitude of the Sun on day 200.

8. Would the shape and size of your graph change if the gnomon had been twice as tall? Explain.

■ Formulating an Equation

We started this activity by describing the actual movement of the Sun in the sky. We abstracted that physical reality, first into an array of measured data, then into a table of numbers, then into a graphical picture, and now into a complete mathematical abstraction: an equation. An equation is usually the most precise way to express mathematical relationships among data points and to predict the values of data not yet measured. The graph you drew in the previous part is a sine wave—abbreviated sin—which can be written approximately as an equation:

$$\theta = 23.5 \cdot \sin{(\text{Days})} + 50. \qquad \textbf{(Equation 3-1)}$$

9. Use the **Equation 3-1**, which describes the Sun's apparent seasonal movement, to once again predict the altitude of the Sun on day 200. How does your answer compare to the answer you determined from your graph in Part 7?

Shadowland—The Sequel

All work must be shown to receive credit.

1. _____
_____
_____
_____

2. **Table 3-1** Shadow length measurements

Days	l	tan θ	θ
0	0.84		
45	0.42		
90	0.30		
135	0.42		
180	0.84		
225	1.50		
270	2.00		
315	1.50		
360	0.84		

3. _____
_____
_____
_____

4, 5, 6.

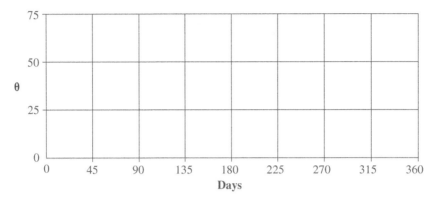

Figure 3-1 Graph of Days vs. θ.

7.

_____ degrees

8. _____

_____

_____

9.

_____ degrees

Shadowland Follow-Up
Homework Assignment

Now that you have completed the activities about the gnomon, apply the general principles of scientific data organization and analysis to a real-world situation, *not in astronomy*. That is, think of an actual situation where observations or experiments could be conducted, the collected data assembled in a table, then pictured as a graph, and finally expressed as an equation. On the worksheet, write a short essay explaining the real-life situation you've chosen. Also detail how you would set up and carry out the data organization and analysis (such as how to arrange your data table and which data to graph), but don't actually perform the analysis. Unlike in-class activities, this is an individual (not a team) assignment. Come up with a unique answer.

Shadowland Follow-Up
Homework Assignment

All work must be shown to receive credit.

_____
_____
_____
_____
_____
_____
_____
_____
_____
_____
_____
_____
_____
_____
_____
_____
_____
_____
_____
_____
_____
_____
_____
_____
_____
_____
_____
_____
_____
_____

The Phases of the Moon

Observing the Moon from night to night, you will notice that different fractions of its visible surface appear illuminated, that is, the Moon goes through phases. Since the source of illumination (the Sun) and the viewpoint (Earth) are fixed relative to one other, the phase of the Moon depends on where the Moon is situated in its month-long orbit around Earth. On the worksheet are several diagrams that will help reveal how the Moon's phases come about.

■ Practice Run

1. The circle at the top of **Figure 5-1** on the worksheet represents a top-down view of a ball that is uniformly illuminated by light coming from the right, as indicated by the arrows. This ball will be viewed from three different positions, represented here by eye-symbols labeled **a**, **b**, and **c**. That is, for each position **a**, **b**, and **c**, you can imagine laying your head sideways on the paper and aligning your eye to the given eye-symbol.

   a. Shade in the portion of the circle at the top of Figure 5-1 that represents the portion of the ball that would remain in *darkness* if illuminated from the right.

   b. Now imagine viewing the illuminated ball at the top of Figure 5-1 from each of three viewpoints, **a**, **b**, and **c**. Envision how the ball would appear to you in each case. In the circle next to each viewpoint, shade in the portion of the ball that would appear *dark*.

■ The Real Thing

2. **Figure 5-2** on the worksheet represents a top-down view of the Moon at a number of locations in its orbit around Earth. (The figure is not to scale.) The Sun's rays illuminate the Moon from the right-hand side of the page, as indicated by the arrows. With the Sun so far away, its rays are virtually parallel when they arrive at Earth and the Moon, as shown.

   a. Shade in the portion of Earth that is *not* illuminated by the Sun. Note: The shaded side corresponds to nighttime on Earth, and the unshaded side to daytime.

   b. For each of the Moon's orbital locations, labeled here **A** through **H**, shade in the portion of the Moon that is *not* illuminated by the Sun. (Do not shade in the Moon's phases here; that comes next.) For positions **A** and **E**, assume that the Moon is not exactly in a line between the Sun and Earth; that is, no eclipse occurs at these positions.

21

   c. In each of the circles **A** through **H** in **Figure 5-3** on the worksheet, shade in the portion of the Moon that would appear dark, as viewed from Earth. The unshaded part within each of these circles represents the visible phase of the Moon: **A**—*new Moon*; **B**—*waxing crescent*; **C**—*first quarter* (commonly called a *half Moon*); **D**—*waxing gibbous*; **E**—*full*; **F**—*waning gibbous*; **G**—*last quarter*; **H**—*waning crescent*.

■ Last Lap

3. In Figure 5-2, Earth is labeled with times of day. Noon occurs on Earth's surface where the Sun appears at its highest daily altitude in the sky, and midnight is the point on Earth that is opposite noon. Sunset, here assumed to take place around 6 pm, occurs at the point where Earth's rotation carries us from our planet's daylight side to its nighttime side; that is, the Sun appears to drop below the horizon. Sunrise, around 6 am, occurs where Earth's rotation brings us back from night into day; here, the Sun appears to rise above the horizon. Studying Figure 5-2, estimate the rising and setting times for the Moon in the phases specified on the worksheet. List the time when the particular phase is first visible from Earth and the time when it is no longer visible from Earth. For example, when the Moon is at point C in its orbit, it rises at around noon and sets around midnight. Hint: When the Moon is visible from Earth, you should be able to draw a line between the two bodies that does not pass through any portion of Earth itself; in other words, Earth does not get in the way of your view of the Moon.

NAME ____________________

WORKSHEET ACTIVITY 5

The Phases of the Moon

All work must be shown to receive credit.

1.

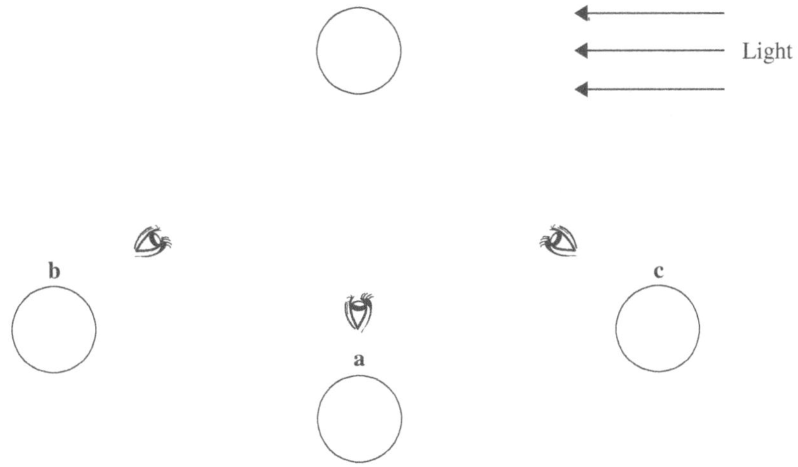

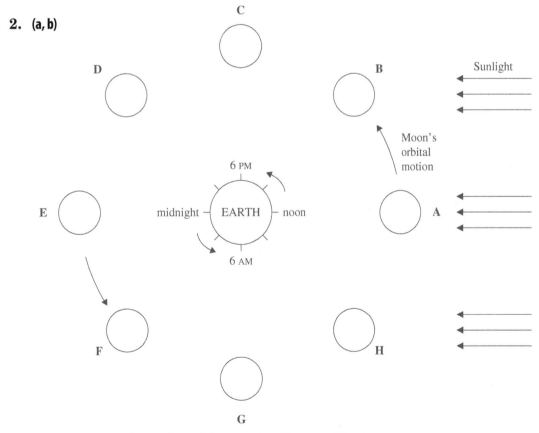

Figure 5-1 Phases of an illuminated ball.

2. (a, b)

Figure 5-2 Top-down view of the Moon's orbit around Earth.

(c)

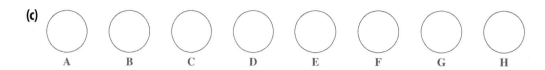

Figure 5-3 The Moon's phases.

3. **B:** rises _____ / sets _____

   E: rises _____ / sets _____

   G: rises _____ / sets _____

Eratosthenes Measures Earth

In trying to understand the cosmos, the ancient Greeks used geometry to dazzling effect. Around 200 BC, the astronomer Eratosthenes conceived a way to measure the size of Earth by simple observation. A brilliant polymath, Eratosthenes had left his native Cyrene, in present-day Libya, to study in Athens before moving to the intellectual hotbed of the Mediterranean: Alexandria, Egypt. He was appointed tutor to King Ptolemy III's son, Philopater, and eventually director of the famous Alexandrian Library.

Eratosthenes learned of a curious phenomenon that occurred once a year in the Egyptian village of Syene (pronounced sī-ē′nē, now the city of Aswan), some 500 miles south of Alexandria. At noon on the first day of summer, the Sun stood at the zenith—directly overhead—in Syene. A gnomon cast no shadow, and the Sun's rays shone straight down to the bottom of a deep well on nearby Elephantine island, illuminating the water below. Eratosthenes was intrigued by this report. He knew that on the same day in Alexandria, the noontime Sun did *not* stand at the zenith, but approximately 7 *degrees* away. In Alexandria, a gnomon cast a shadow, and the Sun's oblique rays struck the wall, not the bottom, of the wells. From this simple observation, Eratosthenes computed the size of Earth, just as you are going to do now.

■ Dividing the Circle

First, some practice with circles and angles. Fact: A circle contains 360 degrees. Dividing a circle will yield pie-shaped sectors whose degree measures must add up to 360 degrees.

1. On the worksheet, draw a line dividing the illustrated circle in half. (See **Figure 6-1**.) What do you think is the degree measure of each sector? Now divide the circle so that it has four equal sectors. What do you think is the degree measure of each of these sectors? Divide the circle further into eight equal sectors. What do you think is the degree measure of each of these sectors?

2. Based on your answers to Part 1, write down a general rule specifying how to compute the degree measure of each sector if given the number of sectors in a circle.

3. Now write down the *reverse* rule, in other words, the rule specifying how to compute the number of sectors in a circle if given the degree measure of each sector.

4. Using your rule from Part 3, compute the number of sectors in a circle if each sector is 7 degrees wide. Round your answer to the nearest whole number. Circle your answer; you'll need it later.

■ Parallel Lines

In deducing the size of Earth, Eratosthenes had to make a critical assumption: that Earth is so small and so far away from the Sun that the Sun's rays are nearly parallel when they strike it. That is, the Sun's rays illuminate Alexandria and Syene from essentially the same direction. You can illustrate that this is true by drawing a scale model of the Earth–Sun system, where the size and spacing of the Sun and Earth have been uniformly shrunk until they "fit" down the side of your worksheet.

- Fact: The Sun's diameter is about 100 times Earth's diameter.
- Earth is roughly 100 Sun-diameters away from the Sun.

5. In the right-hand margin of the worksheet, at the top, draw a *small* circle to represent the Sun. From this "Sun-circle," draw a vertical line along the right-hand side of the worksheet that is 100 times your Sun-circle diameter. Remember, Earth on this scale will be 100 Sun-circles away, so your Sun-circle has to be small enough that 100 of them can fit in a line down the side of the page. Note: You do not need a ruler for this part; do your best to draw a string of 100 Sun-circles of the same diameter.

6. At the lower end of the line you just drew, you will attempt to draw a circle to represent Earth. But first, describe how big your "Earth-circle" would be, *to scale*, compared to your Sun-circle. Your description should be both qualitative and quantitative, that is, in terms of actual numbers and in terms of words. Can you depict the Earth-circle realistically in this reduced-scale drawing?

Look at your scale drawing. Imagine the Sun's rays radiating in all directions from its surface. Now imagine only those rays heading in Earth's direction. Observe how these rays are virtually parallel, regardless of where on Earth they strike. Eratosthenes was right: the Sun's rays illuminate Alexandria and Syene from essentially the same direction.

■ What Eratosthenes Saw

Figure 6-2 on the worksheet shows a cross section of Earth depicting Alexandria, where Eratosthenes lived and, about 500 miles southward, Syene, where the famous well was located. (The well is not drawn to scale, nor is the separation between Alexandria and Syene; both have been exaggerated for clarity.) On this figure, you will draw parallel lines representing the Sun's rays. Note: The rays you will draw here do *not* come from the Sun-circle you drew previously, but originate from an imaginary Sun somewhere far off the page.

7. On the first day of summer, the Sun's rays shone directly down the well in Syene. Draw a line to represent one of these rays, that is, a line that starts near the right-hand edge of the worksheet and extends straight down to the bottom of the well. Extend this line straight to Earth's center, labeled **C** in the figure. Note that this "ray" coincides with the vertical direction in Syene.

8. Draw a second line *parallel* to the line you just drew, again starting near the right-hand edge of the page, but this time extending to the location of Alexandria. Now draw a third line representing the vertical direction in Alexandria; this line should extend from Earth's center **C**, through Alexandria, to a point in space above Earth's surface. The angle between this line and the Sun's ray represents what Eratosthenes saw: in Alexandria, the Sun appeared 7 degrees away from the vertical, even while the Sun was reported to be precisely vertical in Syene. Eratosthenes realized that the difference in the Sun's sky position was due to Earth's curvature. Label the Alexandria angle with its degree measure: 7°.

9. Note the angle that is formed by the intersection of lines at Earth's center **C**. It's no coincidence that this angle is equivalent to the 7-degree angle at Alexandria; there is a geometric rule that explains this. (*When a line crosses a pair of parallel lines, the corresponding interior angles are equal.*) Label the angle at Earth's center with its degree measure: 7°.

■ Finishing Up

Now you can figure out the size of Earth just as Eratosthenes did.

10. Note that the angle you just labeled at Earth's center defines a 7-degree-wide sector of Earth's entire circumference. Each 7-degree sector like this one encloses a 500-mile arc on Earth's surface, the distance between Alexandria and Syene. To compute Earth's circumference, all you need to do is multiply 500 miles by the number of 7-degree sectors in a circle, which you already determined in Part 4. How many miles are in the circumference of Earth? What is the Earth's diameter? (Recall that the circumference of a circle is **π** times the circle's diameter.)

NAME _____

Eratosthenes Measures Earth

For credit, you must show all your work.

1.

Number of Sectors	Degree Measure of Each Sector
2	
4	
8	

Figure 6-1

2. _____

_____

_____

3. _____

_____

4. Number of 7-degree sectors in a circle
(round off to the nearest whole number): _____

5. Complete your scale drawing in the right-hand margin.
(Refer to instructions.)

6. _____

_____

_____

7, 8, 9.

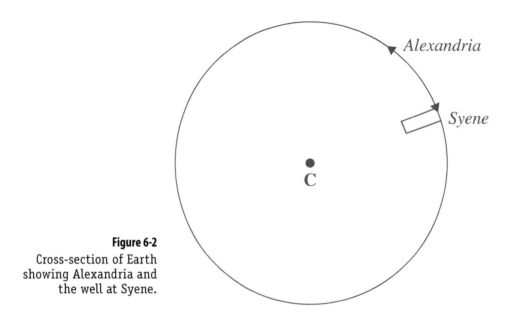

Figure 6-2
Cross-section of Earth
showing Alexandria and
the well at Syene.

10. Circumference of Earth: _____ miles

Diameter of Earth: _____ miles

Aristarchus Measures the Size and Distance of the Moon

In the third century BC, Greek philosopher-mathematician Aristarchus, from the island of Samos in the Aegean Sea, proposed a bold rearrangement of the heavens. For hundreds of years preceding Aristarchus, Greek philosophers believed that Earth occupied the hub of the universe, and that the Sun, Moon, planets, and even the star-studded celestial sphere, which they held enclosed the universe, all circled around it. Aristarchus proposed instead that the *Sun* holds the central position, casting its light symmetrically outward on the other celestial bodies. Ironically, Aristarchus's prime legacy to science turned out to be something other than his Sun-centered universe, which was largely forgotten until Polish mathematician Nicholas Copernicus reintroduced it some 18 centuries later. Aristarchus demonstrated for the first time how it was possible, using simple observations and elementary geometry, to measure sizes and distances of celestial bodies.

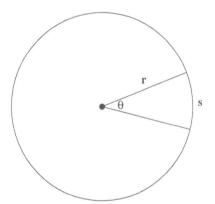

Figure 7-1
Geometry of a sector of a circle.

The starting point for this activity is the geometry of an arc or sector of a circle—basically, a piece of pie. What is the mathematical relationship between the angle enclosed by an arc—its *angular width*—and the length of the arc itself? In **Figure 7-1**, **s** represents the length of the arc; **r** is the radius of the arc; and **θ** (the Greek letter "theta") is the angular width of the arc, that is, how many degrees the arc spans. These quantities are related by the sector equation: $\mathbf{s} = (\mathbf{r\theta})\,/\,57.3$, where **r** and **s** are expressed in units of length (inches, meters, light-years, etc.) and **θ** is expressed in degrees. In astronomical applications, the sector equation can be used to deduce, say, the actual diameter of a celestial object, if its angular diameter and distance are known. Or, if the equation is rewritten as $\mathbf{r} = 57.3\mathbf{s}\,/\,\mathbf{\theta}$, the distance of a celestial object can be computed if the object's true diameter **s** and angular diameter **θ** are known. It's the latter form that Aristarchus used to determine the distance to the Moon.

1. **(a)** Hungry? Here, munch on this 30-degree slice of peach pie, whose radius is 5 inches. What length of pie crust do you get? **(b)** Still hungry? Here's a 30-degree slice of blueberry pie that gives you 4 inches of crust. What is the radius of this pie?

Aristarchus whisked his consciousness far from the surface of our planet and viewed our Earth–Moon system from "above," as though hovering in space. In his geometrical mind, he realized how a lunar eclipse might reveal the Moon's distance from Earth. He already knew the Moon's *angular diameter* in the sky by direct measurement: about a half-degree. So if the eclipse allowed him to determine the Moon's *actual diameter*, he could link these two measurements through the given sector equation and thereby compute the Moon's distance. First, Aristarchus's assumptions:

- The Moon moves around Earth at uniform speed in a perfect circle of radius **r**.
- The Moon takes about 30 days to complete each orbit (called the *orbital period*).
- The Sun is sufficiently far away that its rays are nearly parallel at Earth's surface. Thus, Earth casts into space a cylinder-shaped shadow whose diameter is identical to Earth's diameter **D**, as depicted in **Figure 7-2**. As the Moon moves through Earth's shadow, a lunar eclipse occurs. A photograph of a lunar eclipse in progress is shown in **Figure 7-3**.

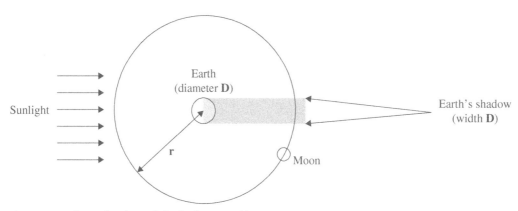

Figure 7-2 Aristarchus's model of a lunar eclipse.

Figure 7-3
Lunar eclipse in progress.
Courtesy of John Walker
http://www.fourmilab.ch/images/
eclipse_lunar_2003_nov/

2. Aristarchus noted that it takes about 3 hours for a point on the Moon—say, a point on its leading edge—to pass entirely through the shadow cast by Earth. He also knew that the Moon takes 30 days (720 hours) to complete an entire orbit, that is, to pass through an angle of 360 degrees. Therefore, at the Moon, the angular width of the eclipse shadow—the "pie crust," in our earlier example—is a small fraction of 360 degrees, equal to $\left(\dfrac{3 \text{ hours}}{720 \text{ hours}}\right) \times 360$ degrees. Compute the angular width of the eclipse shadow in degrees.

3. If the Moon itself appears half a degree wide in the sky ($\boldsymbol{\theta}$), and if the eclipse shadow through which it passes has the width you computed in Part 2, what fraction of the eclipse shadow's width does the Moon occupy?

4. If, as Aristarchus assumed, the eclipse shadow is everywhere as wide as Earth, use your answer to Part 3 to compute the Moon's actual diameter **s**, expressed as a fraction of Earth's diameter **D**. For instance, if the Moon is half as wide as Earth, then its diameter would be written as 0.5**D**.

5. Now that you know the Moon's angular diameter $\boldsymbol{\theta}$ and actual diameter **s**, use the sector equation, $\mathbf{r} = 57.3\mathbf{s}/\boldsymbol{\theta}$, to compute the Moon's distance **r** in terms of Earth's diameter **D**. For instance, if the Moon is 10 Earth-diameters away, its distance is written as 10**D**. It's okay that your answer has the symbol **D** in it; Earth's diameter had not yet been measured in Aristarchus's time, so we'll just leave it as **D**.

6. Draw a scale-model of the Earth–Moon system according to Aristarchus. That is, on the worksheet, draw a circle representing Earth and a second circle representing the Moon. These circles must have the proper relative size and separation, as envisioned by Aristarchus. No ruler is needed. For example, if the Moon were half the Earth's diameter and located 10 Earth-diameters away, your Moon-circle would be half the size of your Earth-circle and would be drawn 10 Earth-circles away.

In fact, Aristarchus overestimated the Moon's diameter; the Moon is actually about *one fourth* Earth's size. And he erred in thinking that Earth's eclipse shadow is shaped like a cylinder; rather, it tapers to a point like a cone. Therefore, Earth's shadow is narrower at the Moon than Aristarchus had assumed. Modern measurements indicate that the Moon's distance **r** is equivalent to about 30 Earth-diameters, or 30**D**.

7. Draw a revised scale-model of the Earth–Moon system, this time according to modern measurements.

Even though Aristarchus turned out to be wrong, his method is sound. A cosmic distance was measured for the first time. But Aristarchus didn't stop there. He reached out further into the cosmos, as you will see in the next activity.

Aristarchus Measures the Size and Distance of the Moon

For credit, you must show all your work.

1.

(a) _____   **(b)** _____

2.

_____ degrees

3. _____

4.

Moon's actual diameter **s**: _____

5.

Moon's distance **r**: _____

6.

7.

ACTIVITY 8

Aristarchus Measures the Size and Distance of the Sun

Following his determination of the Moon's distance and size, Aristarchus applied the forces of geometry, logic, and observation to the Sun. He envisioned how Earth, the Moon and the Sun must be arranged in space at the instant when the Moon appears precisely half-illuminated—in its first-quarter or last-quarter phase. A photograph of the quarter-phase Moon is shown in **Figure 8-1**. At that time, Earth, the Moon and the Sun must form a right triangle, as shown in **Figure 8-2**. It's no different than if you viewed an illuminated ball from the side, perpendicular to the light rays illuminating it. In a right triangle, the cosine—abbreviated cos—of an angle is defined as the *length of the side adjacent to the angle* divided by the *length of the hypotenuse*.

Figure 8-1
Quarter-phase Moon.
© UC Regents/Lick Observatory

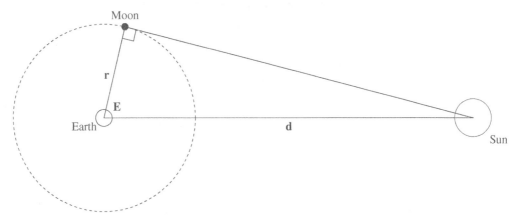

Figure 8-2 Earth–Moon–Sun triangle when the Moon is in the first quarter phase.

For the Earth–Moon–Sun triangle of **Figure 8-2**: $\cos(E) = \frac{r}{d}$, where **E** is the angle between the Moon and Sun as viewed from Earth, **r** is the Moon's distance, and **d** is the Sun's distance. To envision angle **E**, imagine standing outside and extending one arm toward the Sun and your other arm toward the Moon; the angle formed by your outstretched arms is **E**.

A bit of algebra makes the cosine equation look like this: $d = r/\cos(E)$. Having already determined the Moon's distance, all Aristarchus needed to do to find the Sun's distance was measure the angle **E** during the Moon's quarter phase. Easier said than done! If Figure 8-2 were drawn to actual scale, with the Sun very, very far away—that is, **d** much greater than **r**—angle **E** would be very close to, although not quite, a right angle. (In fact, trigonometry had not yet been invented in Aristarchus's time; he used analogous geometrical methods to accomplish the procedure described here.)

1. Aristarchus estimated (guessed?) that during the quarter-Moon phase the angle **E** was 87 degrees. Use the cosine equation in the previous paragraph to determine how many times farther away the Sun is than the Moon. Your answer should take the form $d = n \times r$, where **n** is a number. For now, leave **r** unspecified; it's merely the symbol that represents the Moon's distance. For example, if angle **E** were 30 degrees, then $\cos(E) = 0.866$, and $d = 1.15 \times r$.

2. The solar distance **d** derived from Aristarchus's method is exquisitely sensitive to the adopted value for angle **E**. In other words, even a slight change in angle **E** produces a whopping change in the solar distance **d**. To illustrate, recompute the solar distance if Aristarchus had assumed an angle merely one degree larger, that is, 88 degrees instead of 87 degrees. Again express your answer in the form $d = n \times r$.

3. Modern measurement reveals that Aristarchus was way off in his estimation of angle **E** (so far off, in fact, that we suspect he just plucked a value "out of the hat"). The true value of **E** is 89.85 degrees. Recompute the solar distance now, again in the form $d = n \times r$.

4. Now it should be easy to find the true diameter of the Sun in units of Moon-diameters, in other words, how many Moons would fit across the face of the Sun? The Moon almost precisely covers the Sun during a solar eclipse—the Moon and the Sun *appear* the same angular diameter in the sky. But the Sun is much farther away than the Moon; to *appear* to be the same diameter as the Moon, it must, in fact, be a much larger body than the Moon. For example, if the Sun were 3 times farther than the Moon, yet they *appear* to be the same diameter, the Sun must really be 3 Moon-diameters wide. Or in general, if the Sun is **n** times farther than the Moon, yet they *appear* to be the same diameter, the Sun must be **n** Moon-diameters wide.

    a. Using your value for **n** from Part 1, write down an expression for the Sun's diameter in units of Moon-diameters, according to Aristarchus.

    b. Aristarchus went on to find the Sun's diameter in units of Earth-diameters. To do this, he used his estimate from lunar eclipse observations that the Moon is one third as wide as Earth. Considering this information and your answer to Part 4(a), write down an expression for the Sun's diameter in units of Earth-diameters, according to Aristarchus.

5. To the layperson, Aristarchus's report on his findings is about as exciting as a pamphlet on mixing cement. Not a syllable is wasted on commentary or personal reflection. Yet we can hardly imagine that Aristarchus was unmoved by the extraordinary result of his calculations. Based on your answers to Part 4, can you explain why Aristarchus might have concluded that the Sun, and not Earth, lay at the center of the cosmos? What other unique and important feature of the Sun might have supported this conclusion?

Curiously, Aristarchus did not carry out the next logical step: finding the Sun's and Moon's *distances* in units of Earth-diameters. With such information, he could have formed a true scale model of the Sun–Earth–Moon system, in the same way that a globe shows a scaled-down version of continents and oceans. The essential question is this: given that the Sun is actually _____ Earth-diameters across and the Moon is _____ Earth-diameters across, how far must each object be from Earth to appear as they do, a *half-degree* across, in the sky? To answer this question, we apply the sector equation from the previous activity: $r = 57.3s / \theta$, where r is the Sun's or Moon's distance, s is the Sun's or Moon's true diameter (expressed in Earth-diameters), and θ is the Sun's or Moon's angular diameter in the sky. For example, if an object is actually 10 Earth-diameters wide and it spans an angle of 4 degrees in the sky, its distance is equivalent to about 143 Earth-diameters.

6. Using the sector equation plus Aristarchus's diameters for the Moon and the Sun, compute the distance of **(a)** the Moon and **(b)** the Sun in units of Earth-diameters, according to Aristarchus.

7. Now we're ready to construct the scale model of the Sun–Earth–Moon system, as it might have been envisioned by Aristarchus more than 2000 years ago. Suppose Earth is represented by a Ping-Pong ball, which is approximately 1 inch across. Using your answers to Parts 4 and 6, determine *to scale* the following measurements:

    a. the diameter (in inches) of a ball representing the Moon

    b. the distance (in feet) of that Moon ball from the center of the Ping-Pong ball "Earth"

    c. the diameter (in inches) of a ball representing the Sun

    d. the distance (in feet) of that Sun ball from the center of the Ping-Pong ball "Earth"

Modern measurements reveal that Aristarchus was not too far off in his determination of the Moon's diameter and distance from Earth. However, he erred significantly in his estimates of the Sun's diameter and distance: the Sun is actually more than 100 Earth-diameters across and 11,000 Earth-diameters distant. There's nothing wrong with Aristarchus's methods; they were just impossible to carry out reliably in his day. Nevertheless, Aristarchus's erroneous solar distance became the de facto standard in the Middle Ages. For the first time, someone had calculated the size and distance of a celestial body using observational data gathered from Earth. In bold strokes, Aristarchus had combined geometry, logic, and observation to effectively free himself of his earthly bonds and to show astronomers that it was possible to "lay down a ruler" in the heavens. He had subjected a small portion of the universe to the most basic kind of scientific scrutiny: measurement. Here lay the genesis of modern astronomy.

Aristarchus Measures the Size and Distance of the Sun

For credit, you must show all your work.

1.

Sun's distance **d**: _____

2.

Sun's distance **d**: _____

3.

Sun's distance **d**: _____

4. **(a)** Sun's diameter = _____ Moon-diameters

(b) Sun's diameter = _____ Earth-diameters

5. _____

_____

_____

6. (a) Moon's distance = _____ Earth-diameters

(b) Sun's distance = _____ Earth-diameters

7.

(a) _____ inches

(b) _____ feet

(c) _____ inches

(d) _____ feet

The Copernican Cosmos

In 1543, Nicolaus Copernicus published a massive tome called *On the Revolution of the Heavenly Spheres*. In this work, Copernicus proposed a radical rearrangement of the heavens. The age-old geocentric model posited that Earth is situated at the center of the cosmos, and that the Sun, Moon, and planets circle around it. Copernicus's competing heliocentric model placed the Sun at the cosmic center, with Earth now in orbit.

The new heliocentric model boasted one feature missing from the established geocentric model: by measuring the geometry of certain planetary configurations, Copernicus could derive the relative spacing of the planets. Copernicus used one procedure for *inferior planets*—those whose orbits lie within Earth's orbit—and a somewhat different procedure for *superior planets*—those whose orbits lie outside Earth's orbit. The first part of this activity focuses on the two inferior planets: Mercury and Venus. The second part of the activity deals with the superior planets that were known in Copernicus's time: Mars, Jupiter, and Saturn.

■ Inferior Planets: Mercury and Venus

Since Mercury and Venus have orbits smaller than Earth's, they never stray very far from the Sun in the sky. Thus they are usually visible either shortly before sunrise or shortly after sunset, although Venus can be seen in broad daylight if you know where to look. Mercury is never more than 28 degrees away from the Sun in the sky, an angle designated its *greatest elongation*. Since Venus's orbit is larger than Mercury's, its greatest elongation is correspondingly larger: around 47 degrees. The geometry of greatest elongation is shown in **Figure 9-1**, where θ is the greatest elongation angle, r is the distance of the planet from the Sun, and 1 AU is the term astronomers use to designate the distance between Earth and the Sun, called the astronomical unit. (Earth is located 1 AU from the Sun. Inferior planets are located less than 1 AU from the Sun, and superior planets more than 1 AU from the Sun.)

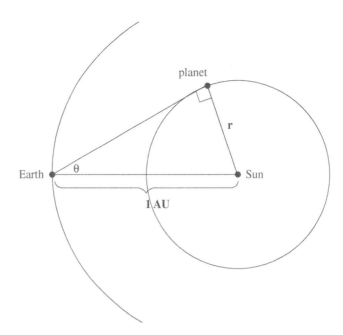

Figure 9-1
Geometry of
greatest elongation.

Copernicus realized that when an inferior planet is at greatest elongation θ, it forms a right triangle with Earth and the Sun, where the right angle is located at the planet. (See Figure 9-1.) This configuration is tailor-made for trigonometry's sine function, defined as *the length of the side opposite an angle* divided by the *length of the hypotenuse*. The definition yields an equation, which can be cast in a variety of different forms as described here:

$$\sin\theta = \frac{r}{1\ \text{AU}} \quad \rightarrow \quad r = (1\ \text{AU})(\sin\theta)$$

or simply

$$r = \sin\theta \quad \text{(in AUs)} \qquad \textbf{(Equation 9-1)}$$

1. By substituting the observed greatest elongations of Mercury and Venus into **Equation 9-1**, compute the distances of these two planets from the Sun, as Copernicus did. Your answers will be in AUs.

2. What is the closest that Mercury and Venus each come to Earth, in AUs? Describe the specific planetary configuration when this occurs. Such a configuration is called *inferior conjunction*.

3. Copernicus also predicted that, in his heliocentric model, Mercury and Venus should each go through phases like our Moon. In what phase would these planets appear when viewed at greatest elongation? Explain. (Hint: Recall how the Moon's phases occur.) Copernicus did not observe these phases, nor could he have—the telescope had not yet been invented. But in the early 1600s, Galileo did observe Venus's phases through a telescope and became a believer in the heliocentric model, in part, because of it.

■ Superior Planets: Mars, Jupiter, Saturn

Since a superior planet's orbit lies outside the Earth's orbit, it forms different configurations with Earth and the Sun than an inferior planet does. (See **Figure 9-2**.) One of these configurations is *opposition*, when the three bodies are aligned in the order Sun-Earth-planet. In the sky, a planet in opposition appears directly opposite the Sun: as the Sun sets below the western horizon, the planet rises above the eastern horizon. A similar alignment, *conjunction*, occurs when the superior planet lies on the other side of the Sun from Earth; here the order is Earth–Sun–planet. The configuration called *quadrature* occurs when the Sun, Earth, and planet together form a right angle. At quadrature, the planet appears 90 degrees away from the Sun in the sky. If you were to extend one arm toward the Sun and your other arm toward the planet, the angle between your arms would reveal the right angle between the two bodies in the sky.

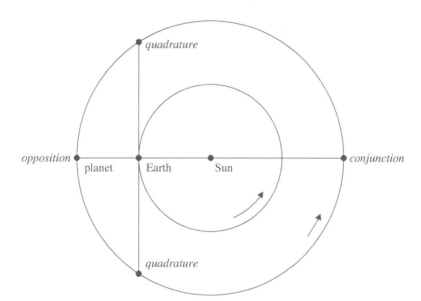

Figure 9-2
Superior planet configurations.

How did Copernicus use these configurations to determine the spacing of the superior planets? For simplicity, let's assume that the planets move in circular orbits at constant speed. (The procedure works for the actual shape of planetary orbits and ellipses, too. It's just more complicated.) Copernicus knew that Earth moves faster in its orbit than the superior planets move in their orbits; thus Earth periodically laps the outer planets, just as a faster race car repeatedly passes a slower one. Copernicus computed Earth's lapping time for the various planets. But he also noted an intriguing geometric relationship in the time interval between two specific configurations: opposition and the quadrature that follows it.

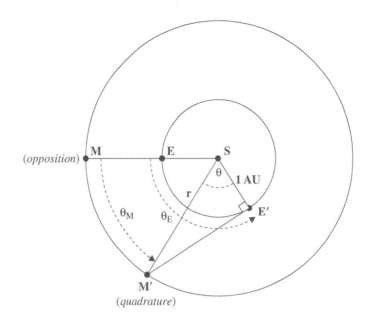

Figure 9-3
Mars at opposition
and quadrature.

Figure 9-3 shows these two configurations for the planet Mars: labels **M** and **E** indicate Mars and Earth at opposition; **M′** and **E′** are the two planets at the quadrature that follows; and **S** is the Sun. Study triangle **SE′M′**. Note that side **SE′** is the standard Earth–Sun distance, 1 AU. Side **SM′** is what Copernicus was seeking: **r**, the distance of Mars from the Sun. If he could only determine the angle labeled **θ**, he could compute **r** by applying the definition of the cosine, or the *length of the side adjacent to the angle* divided by the *length of the hypotenuse* as shown in **Equation 9-2**.

$$\cos\theta = \frac{1\ \text{AU}}{r} \quad \rightarrow \quad r = \frac{(1\ \text{AU})}{(\cos\theta)}$$

Or

$$r = \frac{1}{\cos\theta} \quad (\text{in AUs}) \qquad \textbf{(Equation 9-2)}$$

Copernicus derived the crucial angle **θ** in a clever way. First he figured out how much of an angle Earth moves in its orbit between opposition and quadrature, that is, between points **E** and **E′** in Figure 9-3. Let's call this angle **θ$_E$**. Then he did the same for Mars, this time between **M** and **M′**. Call this angle **θ$_M$**. Figure 9-3 shows that the desired angle **θ** is just the difference between **θ$_E$** and **θ$_M$**, or:

$$\theta = \theta_E - \theta_M \qquad \textbf{(Equation 9-3)}$$

Now let's assemble the outer solar system as Copernicus did.

4. *Computing θ_E.* Copernicus knew that it takes 365 days for Earth to circle the Sun once, that is, for it to move through an angle of 360 degrees. Therefore, each day, Earth moves through an angle given by the mathematical expression $\left(\dfrac{1 \text{ day}}{365 \text{ days}}\right) \times 360$ degrees. Or after **d** days, Earth has moved $\left(\dfrac{d}{365}\right) \times 360$ degrees.

   a. Compute how many degrees Earth moves in its orbit in 1 day.

   b. Copernicus counted how many days elapsed between opposition (point **E**) and quadrature (point **E′**): 107 days. Using the given expression, compute how many degrees the Earth moves in 107 days. This is angle θ_E.

5. *Computing θ_M.* Copernicus also knew that it takes 687 days for Mars to circle the Sun once, that is, for Mars to move through an angle of 360 degrees.

   a. Following the procedure in Part 4, write down a mathematical expression that represents how many degrees Mars moves in its orbit each day.

   b. Now write down an expression that represents how many degrees Mars moves in its orbit in **d** days.

   c. Compute how many degrees Mars moves in the 107 days between opposition and quadrature. This is angle θ_M.

6. *Computing θ.* Substitute your answers to Parts 4 and 5 into **Equation 9-3** to compute the value of angle θ.

7. *Mars's orbit.* Finally, use Equation 9-2 to compute the radius of Mars's orbit, in AUs.

■ EXTRA CREDIT: Jupiter's and Saturn's orbits

Jupiter takes 4332 days—almost 12 years—to orbit the Sun and around 88 days to go from opposition to the next quadrature. Saturn takes 10,761 days—nearly 30 years—to complete its orbit and there are also about 88 days between its opposition and next quadrature.

   a. On a separate sheet of paper, apply the above procedure for Mars to determine the radii of the orbits of Jupiter and Saturn, in AUs.

   b. Using a ruler, draw a careful scale-model showing the spacings of the six known planets in Copernicus's solar system to the scale 1 cm = 1 AU.

The Copernican Cosmos

For credit, you must show all your work.

1.

Mercury: _____ AUs   Venus: _____ AUs

2.

Mercury: _____ AUs   Venus: _____ AUs

_____

3. _____

_____

4.

(a) _____ degrees

(b) $\theta_E = $ _____ degrees

5. (a) _____

(b) _____

(c) $\theta_M = $ _____ degrees

6. $\theta = $ _____ degrees

7. Radius of Mars' orbit = _____ AUs

Kepler's Third Law

In the previous activity, you used geometry to determine the relative spacing of the six planets known in Copernicus's time. In the early 1600s, after Copernicus's heliocentric cosmos had become widely known, the brilliant mathematician Johannes Kepler took the model a step further: he checked whether the radius of each planet's orbit bore any numerical relationship to its orbital period, the time it takes a planet to circle the Sun. The result was Kepler's famous Third Law: the square of the orbital period P is equal to the cube of the orbital radius a, or in equation form, $P^2 = a^3$, where P is expressed in years and a in AUs. Neither Kepler nor his contemporaries knew why planets obey this mathematical rule. His Third Law is empirical: revealed by observation, but whose underlying physics is not known. It would be decades before Isaac Newton uncovered the physical basis for Kepler's law in the fundamental properties of gravity.

1. **Table 10-1** on the worksheet lists modern determinations of the orbital period P and orbital radius a of Mercury, Venus, Earth, Mars, Jupiter, and Saturn. Use your calculator to fill in the rest of the table. Round off your answers to three decimal places.

2. Even though your values for P^2 and a^3 in Part 1 might not be exactly equivalent, do you still feel your results support Kepler's mathematical rule $P^2 = a^3$? Explain the reasoning you used to arrive at your answer.

3. In 1781, in Bath, England, astronomer William Herschel created a worldwide sensation when he discovered a new planet, Uranus. Then in 1846, German astronomer Johann Galle found an eighth planet, Neptune. Subsequent measurements fixed the orbital radii of Uranus at 19.191 AUs and Neptune at 30.069 AUs. Use Kepler's Third Law to compute the orbital periods of **(a)** Uranus and **(b)** Neptune in years. Show your work.

4. Consider the orbital radii of all the planets studied in this activity. The values reveal a curious aspect of the planets' spacing: there is a clustering of planets relatively close to the Sun, followed by what appears to be a gap between Mars and the closest of the outer planets, Jupiter. Italian astronomer Giuseppe Piazzi "filled" this gap in 1801 with his discovery of the first minor planet, or *asteroid*, Ceres. The orbital period of Ceres was found to be 4.603 years. Use Kepler's Third Law to compute the orbital radius of Ceres, in AUs. Show your work. (You will need to compute a cube root to get your answer; ask for help if you need it.)

5. In fact, Kepler's Third Law, as we've been using it, is an abbreviated form of the more complete expression $\mathbf{M} \times \mathbf{P}^2 = \mathbf{a}^3$, or $\mathbf{M} = \dfrac{\mathbf{a}^3}{\mathbf{P}^2}$. Here $\mathbf{M}$ is the mass of the central body—the Sun, in our case—expressed in units of *solar masses*. By definition, the Sun has a mass of precisely 1 solar mass. (That's why the mass symbol $\mathbf{M}$ doesn't appear in the version of Kepler's Third Law that we've been using; it's there, but it equals 1, so we don't have to show it.) The incredible thing about Kepler's Third Law is that it applies to orbiting bodies, not just within our solar system, but anywhere in the universe. For example, we can use it to compute the total mass of the binary star known as Cygnus X-1, which consists of an ordinary star plus a collapsed star—a black hole—in orbit around each other. The orbital period of the Cygnus X-1 pair is 5.6 days (0.0153 years) and one estimate of its orbital radius is 0.2 AU. (The black hole is even closer to its partner star than Mercury is to the Sun.) Solve the "expanded" form of Kepler's Third Law to find the combined mass $\mathbf{M}$ of the star and black hole in Cygnus X-1 in units of solar masses. Show your work.

6. Astronomers believe that so-called "supermassive" black holes exist at the centers of most galaxies, including our own. A supermassive black hole would be far larger and heavier than the one in Cygnus X-1, yet, like any black hole, would be invisible by telescope. Nevertheless, Kepler's Third Law gives us a way to estimate its mass. The law tells us that a star orbiting close (but not too close!) to a supermassive black hole would move much faster along its path than a star circling an ordinary black hole. In fact, a star has recently been discovered that orbits the center of our Milky Way galaxy in precisely this fashion. The star is 1000 AU from the center and takes just 15 years to complete its orbit. As in Part 5, solve the expanded form of Kepler's Third Law to find the combined mass $\mathbf{M}$ of the star and black hole in units of solar masses. Note: The star's mass has been estimated at a "mere" 15 solar masses; therefore, your answer for $\mathbf{M}$ is effectively the mass of the black hole itself. You can see why it's called a supermassive black hole!

Kepler's Third Law

For credit, you must show all your work.

1.

Table 10-1 Kepler's Third Law Data Table

Planet	P (years)	a (AU)	P²	a³
Mercury	0.241	0.387		
Venus	0.615	0.723		
Earth	1.000	1.000		
Mars	1.881	1.524		
Jupiter	11.862	5.203		
Saturn	29.458	9.537		

2. _____
_____
_____

3.

(a) Uranus: _____ years   **(b)** Neptune: _____ years

4.

_____ AUs

5.

_____ solar masses

6.

_____ solar masses

Isaac Newton and the Moon

The hallmark of a successful scientific theory is how well its predictions stack up against the results of precise observation and experiment. Starting in the mid-1600s, English scientist Isaac Newton considered the nature of the force that holds the solar system together. How is it that the planets hurtle through space at breakneck speed, yet remain invisibly anchored to the Sun? What is the origin of Kepler's mathematical laws involving the orbits of planets? As Isaac Newton conceived it, the answer to both questions is a force at once familiar and utterly mysterious: gravity. Gravity anchors us to Earth and plucks an apple from a tree. The very same gravity, Newton insisted, also holds the Moon in its monthly circuit around Earth.

Newton derived a general mathematical law that describes the behavior of gravity. To prove that his proposed law of gravitation was indeed *universal*—it applied not just to objects on Earth, but to the gravitational tug between celestial bodies—Newton compared two numbers: (a) the Moon's orbital acceleration, as predicted by his own gravitational theory; and (b) the Moon's orbital acceleration, this time derived from actual observations of the Moon's movement. If his theory was correct, Newton supposed, these two independent measures of the Moon's acceleration should agree. Before we examine this further, first, a brief physics lesson....

■ Basics of Acceleration

Near Earth's surface, every freely falling object speeds up at the same rate: about 10 meters per second in velocity for every second the object is falling—abbreviated 10 meters/sec^2. In other words, a dropped object has a velocity of about 10 meters/sec after the first second, 20 meters/sec after the next second, 30 meters/sec the second after that, and so on, until it hits the ground. Physicists call this uniform increase in velocity the object's *gravitational acceleration*.

The force that accelerates a freely falling object downward is the mutual gravitational pull between Earth and the object. (We ignore the effects of air resistance which, in reality, make falling feathers accelerate slower than falling rocks.) Away from Earth's surface, where Earth's gravity is weaker, the gravitational acceleration of freely falling objects is less than 10 meters/sec^2. Here's Isaac Newton's question: What is Earth's gravitational acceleration very far away from our planet's surface, say, out at the Moon? And here's Newton's answer.

■ Predicted Acceleration at the Moon

The gravitational force between Earth and the Moon is given by Newton's Universal Law of Gravitation:

$$F_g = \frac{GM_E M_M}{r^2}$$ **(Equation 11-1)**

where M_E is the mass of Earth, M_M is the mass of the Moon, and r is the separation between the centers (not the surfaces) of Earth and the Moon. Newton's Second Law of Motion indicates the acceleration a an object will acquire if subjected to a force F: $a = F/m$, where m is the object's mass. The Second Law can be applied here to the Moon:

$$a = \frac{F_g}{M_M}$$ **(Equation 11-2)**

Substituting the expression for the force F_g from **Equation 11-1** into **Equation 11-2** gives the gravitational acceleration of Earth at the Moon (see how the Moon's mass cancels out from the equation):

$$a = \left[\frac{GM_E M_M}{r^2} \right] \Big/ M_M \quad \rightarrow \quad a = \frac{GM_E}{r^2}$$ **(Equation 11-3)**

Equations 11-1 and **11-3** indicate that, in mathematical terms, both gravitational force and gravitational acceleration diminish with the *square* of the separation r. Thus, Newton reasoned, if the gravitational acceleration a is about 10 meters/sec^2 at Earth's surface, then at 2 times Earth's radius, the acceleration must be $\frac{1}{2^2} \times 10$ meters/sec^2; at 3 times Earth's radius, the acceleration is $\frac{1}{3^2} \times 10$ meters/sec^2; and so on.

1. In Newton's time, the distance to the Moon's center was well established at about 60 times Earth's radius. Based on the preceding examples, compute Earth's gravitational acceleration a at the Moon's center. (It's true that the acceleration will be a bit larger on the near side of the Moon and a bit smaller on the far side, but we'll ignore that.) Express your answer to three decimal places.

Your answer to Part 1 is the *predicted* acceleration at the Moon according to Newton's theory of gravity. Newton needed to compare this prediction to the *actual* acceleration computed from direct observations of the Moon's movement. Time for another physics lesson....

■ Acceleration in Orbit

From basic physics, the acceleration of any object moving in a circle can be computed from the object's velocity v, in meters/second, and the radius r of the circle, in meters, according to the following equation:

$$a = \frac{v^2}{r}$$ **(Equation 11-4)**

Newton's plan was to substitute into **Equation 11-4** the Moon's velocity for **v** and the radius of the Moon's orbit for **r**, then compare the resultant acceleration **a** to that predicted by his theory. But he couldn't directly measure the Moon's velocity. Instead he estimated the velocity from quantities he could measure.

The velocity of an object is a measure of how much *distance* the object travels during a specified *time interval*. Newton knew that the Moon completes an orbit every 27.32 days. Dividing the distance the Moon travels (the circumference of its orbit) by the time interval to cover that distance (the Moon's orbital period) yields the very quantity Newton sought to "plug into" Equation 11-4: the Moon's velocity. Now you do it.

2. Compute the circumference **C** of the Moon's orbit, where $C = 2\pi r$. The radius **r** of the Moon's orbit is about 3.84×10^8 meters. (Scientists routinely use powers-of-ten notation to write large numbers. And there are plenty of large numbers in astronomy! Check the Appendix for a tutorial on powers-of-ten notation.)

3. Convert the orbital period 27.32 days into units of seconds.

4. Divide the orbital circumference from Part 2 by the orbital period from Part 3. Your result is the Moon's velocity **v** in the required units of meters per second.

5. In Equation 11-4, substitute the velocity **v** from Part 4 and the radius **r** of the Moon's orbit given previously. This is the actual acceleration of the Moon.

■ Was Newton Right?

6. If you performed the calculations correctly, the predicted acceleration from Newton's theory of gravity in Part 1 should be the same as the actual acceleration from Part 5. Do these values agree? If they do not agree *exactly*, does that necessarily imply that Newton's theory of gravity is wrong? Explain.

From the result of this analysis and others, Isaac Newton was confident that his law of gravitation was indeed universal—that Earth's gravity holds the Moon in its orbit and, by extension, that the Sun's gravity holds the planets in their orbits. The heliocentric cosmos of Copernicus and Kepler had now truly come of age.

■ EXTRA CREDIT (for the truly adventurous)

Repeat Parts 1 through 6 to confirm that the Sun's gravity holds Earth in its orbit. In other words, compute the Sun's gravitational acceleration at Earth and compare that number to Earth's actual acceleration based on its observed motion. Here are some givens to use in your calculations:

- The gravitational acceleration at the Sun's surface is about 28 times greater than Earth's surface acceleration: 280 meters/sec².
- The radius of Earth's orbit is about 217 times the radius of the Sun itself.
- The radius of Earth's orbit (1 AU) is approximately 1.5×10^{11} meters.
- There are about 3.15×10^7 seconds in a year.

You must show your work neatly for all parts; no extra credit is given for answers only.

Isaac Newton and the Moon

For credit, you must show all your work.

1.

Predicted acceleration **a** = _____ meters/sec^2

2.

C = _____ meters
(in powers-of-ten notation)

3.

Orbital period = _____ sec
(in powers-of-ten notation)

4.

$$\mathbf{v} = \underline{\hspace{2cm}} \text{ meters/sec}$$

5.

$$\text{actual acceleration } \mathbf{a} = \underline{\hspace{2cm}} \text{ meters/sec}^2$$

6.

Galileo Measures a Mountain—On the Moon!

In 1610, Italian astronomer Galileo Galilei created a sensation with the publication of *The Starry Messenger*, an account of his many celestial discoveries with the telescope. Among his observations was a survey of features on the surface of the Moon. He believed that the Moon was another world, like Earth, even comparing mountainous lunar regions to similar landscapes in his native Europe.

One sight in particular captured Galileo's attention, as he relates in his book and its accompanying sketches (**Figure 12-1**):

> *[Not] only are the boundaries of light and shadow in the Moon seen to be uneven and sinuous, but—and this produces still greater astonishment—there appear very many bright points within the darkened portion of the Moon, altogether divided and broken off from the illuminated tract.... Now, is it not the case on the Earth before sunrise, that while the level plain is still in shadow, the peaks of the most lofty mountains are illuminated by the Sun's rays?*

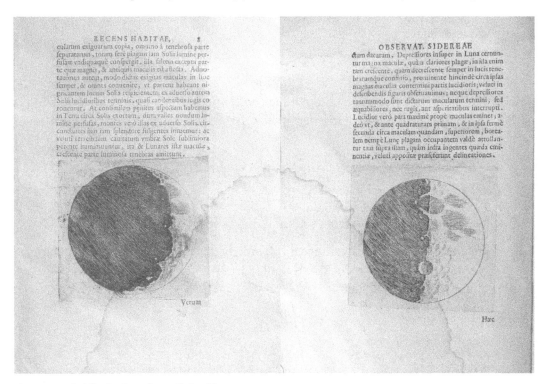

Figure 12-1 Galileo's sketches of the Moon.

Images courtesy History of Science Collections, University of Oklahoma Libraries; copyright the Board of Regents of the University of Oklahoma.

In other words, Galileo believed he saw mountain peaks in the dark portion of the Moon that were so tall that their tops were illuminated by the Sun. In this activity, you will follow Galileo's own simple geometric method to estimate the height of these lunar mountains.

The right-hand image in Figure 12-1 shows Galileo's sketch of the first-quarter Moon, with sunlight shining on it from the right. Several illuminated mountain peaks appear as white specks within the Moon's dark half. Now imagine rotating the Moon, as depicted here, around a horizontal axis until we see one of these illuminated peaks from the side. Then the Moon would appear as Galileo drew it in **Figure 12-2**, with the right-hand hemisphere in sunlight, the left-hand hemisphere in darkness. The base of the mountain is located at point **A** and the top at point **D**. Use your imagination or your pencil to fill in the form of the mountain itself. Figure 12-2 is a hypothetical "sideways" view of the Moon from somewhere in outer space. Our actual "face-on" viewpoint from Earth is indicated by the vertical arrow labeled "our view."

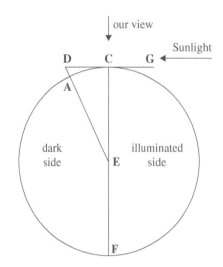

Figure 12-2
Geometry of
an illuminated
mountain.

1. Use the letter labels in Figure 12-2 to fill in the blanks below and on your worksheet. Points are indicated by a single letter, line segments always by a pair of letters, and angles always by three letters.

    a. The letter _____ indicates the Moon's center.

    b. Segments _____, _____, and _____ are each a radius of the Moon.

    c. Segment _____ represents a mountain whose top is at point **D**.

    d. Segment _____ indicates the beam of sunlight that touches, at point **C**, the boundary between the Moon's light and dark halves—technically, the *terminator*—and illuminates the mountain.

2. In Figure 12-2, Galileo found a right triangle of interest: triangle **ECD**. Again, use the letter labels in Galileo's diagram to fill in the blanks below regarding statements about triangle **ECD**.

    a. Angle _____ is the right angle.

    b. Side _____ is a lunar radius, about 1080 miles.

    c. Side _____ is the distance from the terminator to the illuminated mountain peak, as seen from Earth.

    d. Side _____ is the triangle's hypotenuse.

    e. This hypotenuse encompasses the sum of two segments: another lunar radius, segment _____, plus the height of the mountain, segment _____.

3. In *The Starry Messenger*, Galileo wrote that the illuminated peaks appeared as far as $\frac{1}{20}$ of the Moon's diameter from the terminator. **(a)** Given that the Moon's diameter is 2160 miles, compute how far into the Moon's dark side these illuminated peaks were situated, according to Galileo. **(b)** To which side of the right triangle **ECD** does this answer correspond?

4. Galileo now applied the Pythagorean Theorem for right triangles: the square of the hypotenuse is equal to the sum of the squares of the sides, or $(\text{hypotenuse})^2 = (\text{side a})^2 + (\text{side b})^2$. Using your answers to Parts 2(b), 2(c) and 2(d), write down the Pythagorean Theorem for the right triangle **ECD**.

5. Using the numerical values from Parts 2(b) and 3, solve the Pythagorean Theorem equation for the hypotenuse of triangle **ECD**.

6. Since the hypotenuse encompasses the sum of a lunar radius plus the height of the mountain, subtract a lunar radius from your answer to Part 5 to obtain the height of the mountain itself. In fact, your answer represents the minimum height of lunar mountains; they must be *at least* this tall to poke up into the sunlight.

7. **(a)** Compare the height of lunar mountains to that of mountains on Earth. Your answer must include a direct number comparison. If you don't know how high mountains on Earth are, ask someone or look it up. **(b)** Do you agree with Galileo's stated conclusion that lunar mountains are *several times* taller than the highest mountains on Earth?

8. Having found that the Moon is so rugged, Galileo had to explain why the edge, or *limb*, of the Moon nonetheless appears round and smooth instead of jagged. He proposed that, when looking at the limb of the Moon, an observer sees successively more distant mountains in the gaps between nearer mountains. Explain how this might make the Moon's limb appear round and smooth. Feel free to sketch a picture to support your explanation. (Galileo also suggested that the Moon's atmosphere blurs the ruggedness of the landscape, making it look more regular than it is. In this he was wrong; the Moon has no atmosphere.)

Galileo Measures a
Mountain—On the Moon!

For credit, you must show all your work.

1. **(a)** _____ **(b)** _____, _____, _____ **(c)** _____ **(d)** _____

2. **(a)** _____ **(b)** _____ **(c)** _____ **(d)** _____ **(e)** _____, _____

3. **(a)** _____ miles **(b)** _____

4. _____

5.

_____ miles

6.

_____ miles

7. (a) _____

_____

(b) _____

_____

8. _____

_____

_____

_____

Precision Astronomy After Galileo—Stellar Aberration

The crude telescopes of Galileo's era evolved during the 1700s and early 1800s into precision measuring instruments. These new-generation telescopes had high-quality lenses, sturdy mounts, and, sometimes, mechanisms that allowed them to track the movements of stars across the sky.

One of the astronomical imperatives that drove the improvement of telescopes was the desire to validate Copernicus's heliocentric cosmos by observing the elusive *stellar parallax*. According to the Copernican model, stars should appear to "wobble" slightly in their places, in response to Earth's wide-swinging motion around the Sun. And although by the 1700s most scientists believed that Earth indeed circles the Sun, they still had no direct observational evidence that this is true. Whoever was first to detect the parallax shift of stars would go down in history as the one who proved the Copernican cosmos. But as is so often true in science, when searching for one phenomenon, you may wind up discovering another.

Among the astronomers who tried to detect stellar parallax was James Bradley, a clergyman in the 1720s and, later, England's Astronomer Royal. Through careful observation, Bradley found that the star Gamma Draconis executes an annual wobble of approximately six thousandths of a degree—0.006 degree—from its average position in the sky. By comparison, the angle spanned by the full Moon is around half a degree, more than 80 times larger.

As small as 0.006 degree might seem, Bradley decided that it was too large to be the long-sought parallax effect. A 0.006-degree parallax would have implied that Gamma Draconis lay much closer to Earth than stars were generally believed to be. Also, Gamma Draconis's shift was always in the same direction that Earth was moving at the time; the shift clearly arose from Earth's *velocity*, not from its changing *position* around the Sun. Bradley concluded that he had discovered a completely new phenomenon. Equally exciting, the unexpected oscillation of stars due to *stellar aberration*, as it came to be called, indicated that Earth revolves around the Sun. Seeking stellar parallax, Bradley had stumbled across a completely different proof of the heliocentric cosmos.

Stellar aberration has a number of everyday analogues. For example, when running through a rain shower, a person has to tilt an umbrella forward to keep from getting wet, as though the rain were falling, not straight down, but at an angle. Or when driving through the rain, the watery streaks on a car's side window are not vertical, but angled. In both cases, the horizontal velocity of the observer combines with the vertical velocity of the rain to create the illusion of an angled rainfall. Similarly, the velocity of Earth in its orbit around the Sun combines with the velocity of starlight to create the illusion that light from a star enters a telescope at an angle—that the star has shifted from its true position.

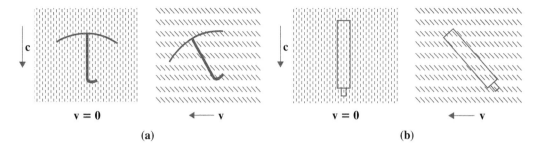

Figure 13-1 Aberration of raindrops and starlight.

Figure 13-1 illustrates the geometry of two situations: (a) vertical raindrops falling at velocity **c**, first in the case of a stationary umbrella, then in the case of an umbrella being whisked sideways at velocity **v**; and (b) "vertical" starlight moving at velocity **c** (the speed of light), first in the case of a stationary telescope, then in the case of a telescope moving "horizontally" at velocity **v**. In reality, all earthbound telescopes are moving because Earth itself is moving along its orbit.

The angle at which the umbrella in Figure 13-1(a) must be tilted—that is, the angle at which the raindrops *appear* to fall—depends on how large the umbrella velocity **v** is compared to the raindrop velocity **c**. Imagine a right triangle, as in **Figure 13-2**, whose vertical side is a line segment representing the speed **c** of the raindrops, and whose horizontal side is a line segment representing the speed **v** of the person carrying the umbrella. The hypotenuse of this triangle conveniently indicates the angle **θ**, relative to the vertical from which the rain *appears* to be coming—or, equivalently, it indicates the tilt the umbrella must have to most effectively intercept the rain.

For any right triangle, the tangent of an angle is defined as the *length of the the side opposite the angle* divided by the *length of the side adjacent to the angle*. For the right triangle in Figure 13-2 the following equation applies:

$$\tan \theta = \frac{v}{c} \qquad \text{(Equation 13-1)}$$

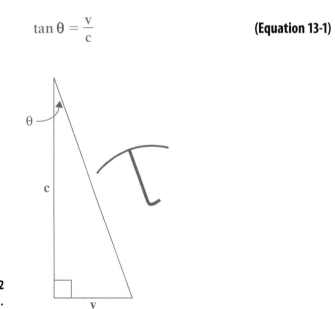

Figure 13-2
Aberration of falling raindrops.

1. Suppose raindrops are falling straight down with a speed of 15 feet per second. Use **Equation 13-1** to compute the tilt of the umbrella when the person carrying the umbrella moves at **(a)** 5 feet per second, **(b)** 10 feet per second, and **(c)** 15 feet per second.

2. Use your answers from Part 1 to write a statement describing how the tilt of the umbrella depends on the raindrops' velocity and the velocity of the person carrying the umbrella.

3. Suppose raindrops are falling straight down with a speed of 15 feet per second. This time you observe a person walking through the rain with an umbrella tilted at an angle of 10 degrees. Use Equation 13-1 to compute the velocity of the person.

Figure 13-1(b), which illustrates the analogous situation for starlight entering an Earth-based telescope, suggests that we can apply the trigonometric principle described in Equation 13-1 to the phenomenon of stellar aberration—and thereby compute the orbital velocity of Earth. James Bradley imagined the right triangle in **Figure 13-3** formed by line segments representing the velocity of light **c** (the longer side) and the orbital velocity of Earth **v** (the shorter side). The triangle's hypotenuse indicates the direction Bradley had to point his telescope to intercept starlight from Gamma Draconis. (The triangle is not drawn to scale.) Figure 13-3 suggests that were Earth moving close to the speed of light, the majority of stars would appear huddled in one portion of the sky!

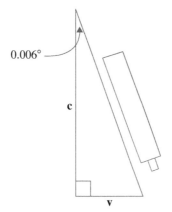

Figure 13-3
Aberration angle of starlight.

4. Given Earth's motion, Bradley had to tilt his telescope 0.006 degree forward from Gamma Draconis's true position in order to sight the star. Use Equation 13-1 to compute Earth's orbital velocity in miles per second. Note: The speed of light **c** is 186,000 miles per second.

5. In science, it's always a good idea to check your answer to a particular problem against one derived by a different method. For instance, to check your answer to Part 4, you can also find Earth's orbital velocity by applying the basic definition of velocity: velocity = distance/time. That is, compute Earth's orbital velocity by dividing the *circumference of Earth's orbit* by the *number of seconds it takes Earth to complete the orbit*. Here's the relevant data: the radius of Earth's orbit is 93 million miles (9.3×10^7 miles) and 1 year consists of 3.15×10^7 seconds. The circumference of a circle is given by the formula $C = 2\pi r$, where **r** is the circle's radius.

6. Your answers from Parts 4 and 5 for Earth's orbital velocity may differ. Comment on why that might be the case. Be specific.

Precision Astronomy After Galileo—Stellar Aberration

For credit, you must show all your work.

1.

(a) _____ degrees

(b) _____ degrees

(c) _____ degrees

2. _____

_____

_____

3.

_____ feet per second

4.

_____ miles per second

5.

_____ miles per second

6. _____

_____

_____

Precision Astronomy After Galileo—Stellar Parallax

By the 1820s, a century after James Bradley accidentally discovered stellar aberration, telescopes had improved sufficiently that detection of the elusive stellar parallax seemed within the astronomer's grasp. Yet it was no longer necessary to detect the tiny parallax shift of a star—any star!—to prove that Earth circles the Sun; stellar aberration had already done that. Nineteenth-century astronomers strove to measure stellar parallaxes for a wholly different reason: to gauge the scale of the cosmos. Stellar parallax provides a direct geometric means to measure the distances of stars—the smaller a star's parallax, the farther away the star.

It was not until 1838 that German astronomer-mathematician Friedrich Bessel succeeded in measuring the first parallax of a star. As expected, it was exceedingly small, confirming what astronomers had already suspected—the distance of stars is truly vast. Today, the parallaxes, and corresponding distances, of millions of stars have been measured. This activity will acquaint you with the basics of distance estimation using parallax measurement.

■ Wobbly Star

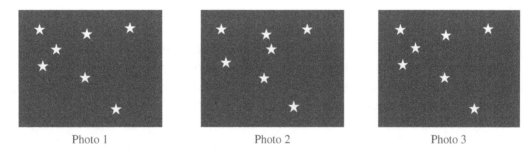

Photo 1   Photo 2   Photo 3

Figure 14-1 Simulated images of a field of stars.

The illustration (**Figure 14-1**) shows a series of simulated "photographs" of the same field of stars.

- *Photo 1:* Stars as they appeared one night in the sky

- *Photo 2:* Same stars as they appeared 6 months later

- *Photo 3:* Same stars 6 months after Photo 2 (1 year after the original photograph, Photo 1)

Although the stars may look alike (real stars all appear as bright points, even through a powerful telescope), one of these stars is actually much closer to Earth than the others. You can determine which star is the close one by recalling how stellar parallax relates to distance, this time in the reverse sense: the larger a star's parallax, the closer the star.

1. How does Earth's position around the Sun differ from Photo 1 to Photo 2 to Photo 3?

2. How can you tell which star in the photographs is closest to Earth?

■ How Far Is Your Finger?

Your instructor will have arrayed horizontally on the blackboard a series of marks spaced ½ foot apart and numbered 0.0, 0.5, 1.0, 1.5, and so on. In this part of the activity, your finger represents a nearby star, the marks on the blackboard represent distant stars, and your eyes represent the Earth at two opposite locations in its orbit around the Sun.

3. Hold your finger upright in front of your eyes about 8 inches away (roughly the width of a common sheet of paper). **(a)** Close one eye. Line up your finger with one of the marks on the blackboard. Have your partner record the number of the lined-up mark. **(b)** *Without moving your finger*, sight your finger instead with the eye that had been closed. Have your partner record the number of the new lined-up mark. **(c)** Subtract the two mark-numbers. The answer is the parallax shift θ of your nearby finger in fictitious units we'll call "marks."

4. Now move your finger an arm's length away from your eyes. Repeat the procedure in Part 3. This time, the answer to Part (c) is a measure of the parallax shift θ of your outstretched finger, again in units of marks.

5. Which has the larger parallax, your nearby finger or your outstretched finger? Is this what you would expect? Explain.

6. Just as astronomers use a star's parallax shift to compute the distance to the star, you can use your finger's parallax shift to compute the distance to your outstretched finger. But first you have to convert your finger's parallax shift from the fictitious unit of marks to the standard unit of angles: degrees. This requires two steps: first take your answer for θ from Part 4(c) and multiply it by 57.3; then divide the result by the distance between you and the blackboard, in feet. (Either you or your instructor will carry out this measurement in the classroom.) In equation form, the procedure is:

$$\theta_{\text{degrees}} = \frac{(\theta_{\text{marks}} \text{ for your outstretched finger}) \times (57.3)}{(\text{distance from you to the blackboard, in feet})}$$

7. Now you're ready to use this parallax angle in degrees to compute the distance to your outstretched finger. **Figure 14-2** illustrates a top-down view of the parallax experiment. Notice that the intersecting lines-of-sight from your eyes form two triangular sectors: one sector formed by your eyes and your finger and the other sector formed by your finger and the marks on the blackboard. Recall the sector equation from Activity 7, with which Aristarchus measured the distance and size of the Moon: $r = 57.3s / \theta$. We can make good use of it here. In the left-hand sector of Figure 14-2, r is distance of your outstretched finger; s is the separation between

your eyes (let's assume it's 2.5 inches); and θ is the parallax angle you obtained in Part 6. Use the sector equation to compute the distance to your outstretched finger, in inches. Is your answer reasonably close to what you think it should be?

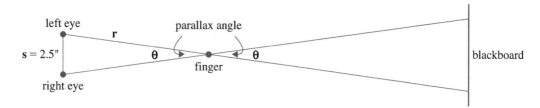

Figure 14-2 Parallax of your finger.

8. Recall that the spacing between your eyes symbolizes the width of Earth's orbit, and your outstretched finger symbolizes a nearby star. In reality, the distance to the *nearest* star beyond the solar system is *150,000* times the width of the Earth's orbit. This implies that, were Figure 14-2 drawn to the proper scale, **r** would be 150,000 times as long as **s**! And your outstretched finger, standing in for the star, would be located at a distance 150,000 times the spacing between your eyes. If the typical spacing between human eyes is 2.5 inches, compute how far away the nearest star would lie on this reduced scale. Express your answer in miles. Show your work. Conversion factors: 1 foot = 12 inches; 1 mile = 5280 feet.

■ The 3D Cosmos

Astronomers have divided the night sky into 88 constellations—arrangements of stars that have been associated with mythological characters, animals, or sometimes everyday objects. While a constellation's stars appear close to one another in the sky, they are often widely separated from each other in space. Once astronomers succeeded in measuring stellar parallaxes, a cosmic third dimension—distance—was added to our age-old, two-dimensional perspective on the night sky. In mapping out the stars near our solar system, astronomers have adopted a distance unit called the *parsec*, which is defined simply as the reciprocal of a star's parallax (**p**):

$$d = \frac{1}{p} \qquad \textbf{(Equation 14-1)}$$

In this equation, the parallax angle must be expressed in units of *arcseconds*, or ¹⁄₃₆₀₀ of a degree. (The astronomer's parallax angle **p** is defined as half of the parallax angle θ you used to determine the distance to your finger. This makes no difference to the work you will do next.)

9. **Table 14-1** on the worksheet lists the parallaxes (in arcseconds) of the seven major stars that define the Big Dipper, a well-known, northern-hemisphere star pattern. Use **Equation 14-1** to compute the distance to each of the stars. Write your answers in the table.

10. Figure 14-3 on the worksheet shows two views of the Big Dipper. On the left appears the flattened "constellation" view that we see in the night sky. To the right appears the space-based view, which will help us visualize the seven stars according to their actual distances from Earth. This latter view will reveal whether the Big Dipper's stars, which appear close to one another in the sky, are, in fact, close to one another in space. For each star in the table, plot a data point along that star's horizontal line at the star's computed distance. After you've plotted all seven stars, fold the worksheet page along the indicated line so that the two parts of the page are perpendicular to each other. Here is a three-dimensional rendering of the Big Dipper's stars!

11. Which of the Big Dipper's stars are grouped together in space? Which are not?

Precision Astronomy After Galileo—Stellar Parallax

For credit, you must show all your work.

Photo 1   Photo 2   Photo 3

Figure 14-1 Simulated images of several stars.

1. _____
   _____

2. _____
   _____

3. (a) _____   (b) _____   (c) θ of nearby finger = _____ marks

4. (a) _____   (b) _____   (c) θ of outstretched finger = _____ marks

5. _____
   _____

6.

$\theta_{degrees}$ = _____ degrees

7.

r = _____ inches

8.

_____ miles

9.   **Table 14-1** Parallaxes of Big Dipper Stars

Star	Parallax (arcseconds)	Distance (parsecs)
Dubhe	0.0026	
Merak	0.0041	
Megrez	0.0040	
Phad	0.0039	
Alioth	0.0040	
Mizar	0.0042	
Alkaid	0.0032	

10.

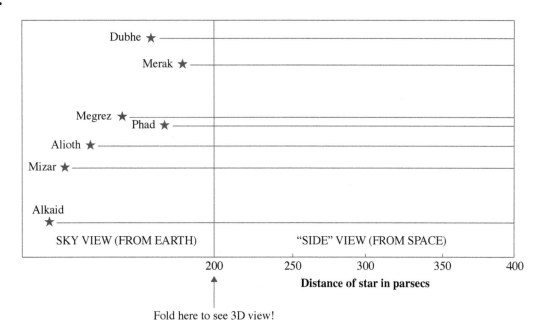

Figure 14-3 Big Dipper: sky and space views.

11. _____
_____
_____

Picturing the Universe— How Photography Revolutionized Astronomy

The invention of practical photography by Louis Daguerre was announced in Paris in 1839. News of the daguerreotype process spread quickly around the world, and shortly the first photographs of the Moon and the Sun were produced by attaching a camera to the eyepiece-end of a telescope. Photographs of fainter celestial objects—planets, stars, nebulae, comets—proved more challenging. To capture their feeble light on a photographic plate required long time-exposures. This, in turn, demanded both speedier photographic processes and, to prevent night-sky targets from drifting out of view, telescope tracking-mechanisms more accurate than what was commonly available in the mid-1800s.

Much of the groundwork for the photographic revolution in astronomy was laid by amateur astronomers. While their professional counterparts focused on the classical astronomy of stellar positions and motions, a succession of dedicated amateurs developed the means to apply both photography and spectroscopy to a host of astrophysical problems: What are stars made of? What are the faint, cloudlike nebulae? What are the conditions on the Sun's surface? Still, few professional astronomers admitted the potential of the new technologies. The turning point came in the 1890s when celestial photographers began to discover objects in space that were too faint to be seen though a telescope by eye alone.

To appreciate how photography helped remake astronomy into a modern science, you are going to give your impressions of a series of astronomical images from various eras. Some are drawings of celestial objects as they appeared by eye through a telescope; others are photographs recorded by attaching a camera to a telescope. But first, here's some practice in the art of observation.

■ Faraday's Face

Figure 15-1 includes several images of the famous 19th-century physicist Michael Faraday, who invented the electric motor and electric generator.

1. Study the pictures carefully and decide which one best represents Faraday's true appearance. Explain why you trust this image the most.

2. Describe some of the similarities and differences between your "most trusted" image and the others.

3. Why do you think there is such a wide variation in Faraday's appearance among the various images?

Figure 15-1 Images of Michael Faraday.

4. Extend your reasoning to predict the relative trustworthiness (accuracy) of drawings of astronomical objects, as they appear to the eye through a telescope, versus photographs of the same objects taken through a telescope.

■ The Moon's Face

Figure 15-2 shows several images of the moon. Figure 15-2A is a drawing of the Moon by Galileo from 1610; Figure 15-2B is a drawing by astronomical artist Etienne Trouvelot from 1875. (Trouvelot is reviled by some for his disastrous introduction of the gypsy moth to the United States.) Figure 15-2C is a vintage lunar photograph taken in 1865 by astronomer Lewis Morris Rutherfurd from his backyard observatory in New York City, and Figure 15-2D is a photograph taken in 1971 by astronaut David Scott on the U.S. Apollo 15 mission to the Moon.

5. Compare the drawings with each other and with the photographs. Describe similarities and differences among them. Assess the relative trustworthiness of the drawings versus the photographs.

■ Jupiter's Face

Figure 15-3 shows Etienne Trouvelot's drawing of Jupiter from the 1870s and a Voyager 1 spacecraft photograph taken about a century later.

6. Compare the drawing and the photograph. Cite similarities and differences of specific features on Jupiter.

■ Diffuse Nebulae: Clouds of Gas in Space

The Orion nebula is a distant interstellar gas cloud that can be seen with the unaided eye as a fuzzy patch below the "belt" of the constellation Orion the hunter. **Figure 15-4** comprises several different images of the nebula. Figure 15-4A is William Lassell's drawing of the nebula from 1852. Figure 15-4B, at a somewhat different scale, is the first-ever photograph of that object, taken through a telescope by New York amateur astronomer Henry Draper in 1880.

7. Compare these two images of the Orion nebula. Describe similarities and differences.

8. Which do you think better reflects the true appearance of the Orion nebula, Lassell's drawing or Draper's photograph? Explain.

9. Given that Draper was a skilled astronomer, why do you think his photo is so poor? Consider that in the 40 years since the advent of photography, no one before Draper had succeeded in taking a picture of any nebula through a telescope.

Figure 15-4C shows a much-improved photograph of the Orion nebula taken by English amateur astronomer A. A. Common in 1883, just 3 years after Henry Draper's pioneering attempt. Figure 15-4D is a photograph of the same object taken by the Hubble Space Telescope in 2006.

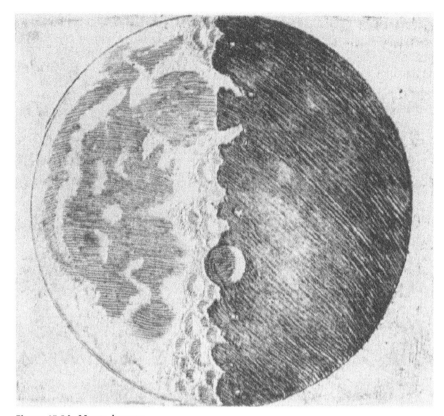

Figure 15-2A Moon image.

Images courtesy History of Science Collections, University of Oklahoma Libraries; copyright the Board of Regents of the University of Oklahoma.

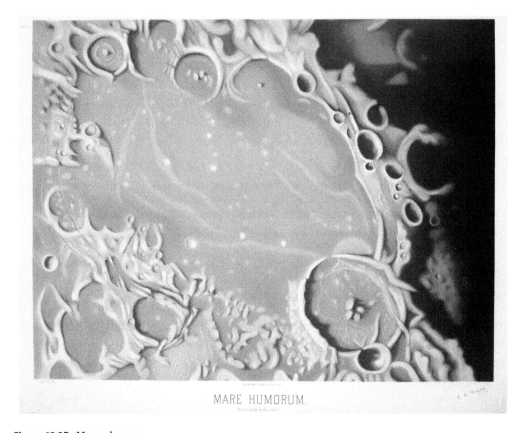

Figure 15-2B Moon image.

Figure 15-2C Moon image.
© CCI Archives/Science Photo Library

Figure 15-2D Moon image.
Courtesy of NASA

THE PLANET JUPITER.

Figure 15-3A Jupiter image.
© *Science, Industry & Business Library, The New York Public Library, Astor, Lenox and Tilden Foundations*

Figure 15-3B Jupiter image.
Courtesy of Voyager 1/NASA

10. Compare these images to Lassell's drawing and to each other. Describe similarities and differences.

11. For his 1883 Orion nebula picture, Common used a telescope with a light-collecting mirror that was more than twice as wide as Draper's. He also adopted a new photographic process that was much more sensitive to light than the one Draper used. Explain how these two factors might have made Common's photograph a better rendering of the Orion nebula than Draper's.

12. The Hubble Space Telescope has a light-collecting mirror much larger than either Draper's or Common's telescope, is equipped with a high-sensitivity digital camera, and is situated in orbit around Earth. Explain how these factors might have made the Hubble picture a better rendering of the Orion nebula than the earlier photographs.

■ A Galaxy Revealed

Among the great astronomical mysteries of the 1800s were the spiral nebulae, which appeared in telescopes as swirling cloudlike forms, but whose nature was unknown. Four different renderings of one such spiral nebula, designated M51, are shown in **Figure 15-5**. Figure 15-5A is a drawing from 1845 by William Parsons, the Earl of Rosse, who viewed M51 through what was then the world's largest telescope and sketched what he saw. Figure 15-5B is a picture from 1889 by English amateur astronomer Isaac Roberts, and is perhaps the first-ever photograph of M51. Figure 15-5C is a 1950 photograph that was taken through the largest-in-the-world telescope of that era, the 200-inch Hale reflector on Mt. Palomar in California. Figure 15-5D was obtained with the Hubble Space Telescope in 2005. Today we know that spiral nebulae like M51 are actually spiral galaxies, vast star systems like our own Milky Way galaxy.

13. Compare the four pictures of M51. Describe similarities and differences.

14. What factors might have led to the noticeable improvements in these images over time?

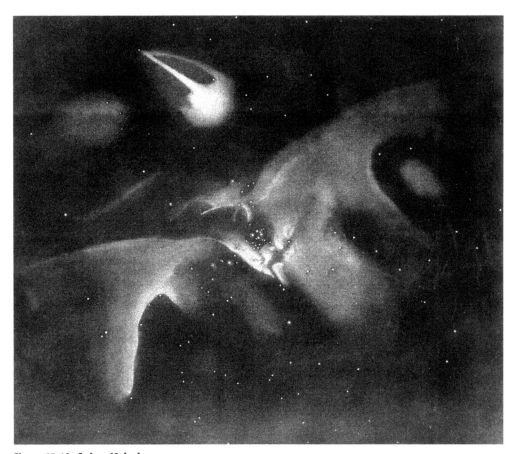

Figure 15-4A Orion Nebula.
© *Royal Astronomical Society/Photo Researchers, Inc.*

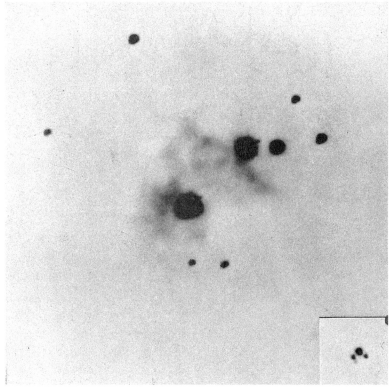

Figure 15-4B
Orion Nebula.
© *Harvard College Observatory*

Figure 15-4C Orion Nebula.
Reproduced from Agnes M. Clerke. A Popular History of Astronomy During the Nineteenth Century.
Adam & Charles Black, 1887.

Figure 15-4D
Orion Nebula.
*Courtesy of NASA/ESA,
M. Robberto (STSi/ESA) and
the Hubble Space Telescope
Orion Treasury Project Team*

Figure 15-5A M51—Whirlpool galaxy.
© *Science Photo Library*

Figure 15-5B
M51—Whirlpool galaxy.
© *Science Museum/Science
& Society Picture Library*

Figure 15-5C M51—Whirlpool galaxy.
Courtesy of the Palomar Observatory and Caltech

Figure 15-5D M51—Whirlpool galaxy.
Courtesy of DSS/NASA/JPL-Caltech

Picturing the Universe—How Photography Revolutionized Astronomy

For credit, you must show all your work.

1. _____
   _____
   _____
   _____

2. _____
   _____
   _____
   _____

3. _____
   _____
   _____
   _____

4. _____
   _____
   _____
   _____

5. _____

_____

_____

_____

_____

_____

_____

6. _____

_____

_____

_____

_____

_____

_____

7. _____

_____

_____

_____

_____

_____

_____

8. _____

_____

_____

_____

_____

_____

_____

9. _____

_____

_____

_____

10. _____
_____
_____
_____
_____
_____
_____

11. _____
_____
_____
_____
_____
_____
_____

12. _____
_____
_____
_____
_____
_____
_____

13. _____
_____
_____
_____
_____
_____

14.

How Bright Is That Star? A Tutorial on the Magnitude System

In describing the stellar and galactic realms of the cosmos, we need a quantitative way to indicate the brightness of stars and other celestial objects. As originally developed by ancient Greek astronomer Hipparchus, the astronomical magnitude system assigns a number to each of 6 different classes of brightness: magnitude **1** to the very brightest stars, magnitude **2** to the next brightest, and so on, up to magnitude **6**, which is assigned to the faintest stars visible to the unaided eye in a very dark sky. In other words, the *brighter* the star, the *smaller* its magnitude number. For example, a star of magnitude **2** is brighter than a star of magnitude **5**. (Astronomers are used to this counterintuitive, "backwards" ranking of brightness.)

Telescopes have revealed celestial objects fainter than the eye can see, that is, objects with magnitude numbers larger than **6**. The stars astronomer Edwin Hubble studied during the 1920s in the faraway Andromeda galaxy have magnitudes of **18** or **19**. Modern telescopes, coupled with exquisitely sensitive electronic light detectors, show even fainter celestial objects, with magnitudes around **30**. At the opposite extreme, very bright objects—the Sun, the Moon, plus a handful of planets and stars—have magnitudes numbers that are **negative**; the more *negative* the magnitude, the *brighter* the object. The Sun's magnitude is **−26.7** and that of Sirius, the brightest star in the sky, is **−1.5**. (By the way, the North Star, at magnitude **2.0**, is *not* the brightest star in the sky.)

Magnitudes are also used to compute how many times brighter one star is than another—a brightness ratio—according to the following rule: each 1-magnitude *difference* between stars represents a brightness *ratio* of about 2.5. Thus, to compute the brightness ratio that corresponds to a magnitude difference of (say) 3, merely multiply 2.5 by itself 3 times: $2.5 \times 2.5 \times 2.5$, which is about 16. Here are several more examples:

- A star of magnitude **5** is 2.5 times brighter than a star of magnitude **6**. Likewise, a star of magnitude **6** is 2.5 times fainter than a star of magnitude **5**.
- A star of magnitude **5** is 2.5×2.5, or about 6.25, times brighter than a star of magnitude **7**.
- A 5-magnitude difference between stars represents a brightness ratio of $2.5 \times 2.5 \times 2.5 \times 2.5 \times 2.5$, that is, $(2.5)^5$, or about 100.

1. **(a)** How many times brighter is a star of magnitude **10** than one of magnitude **13**?
(b) How many times brighter is the planet Mars when it shines at magnitude **−2.0** than the North Star, with magnitude **2.0**?

2. If two stars have magnitudes m_1 and m_2, respectively, a shorthand scheme to convert their magnitude difference $m_1 - m_2$ into a brightness ratio is $(2.5)^{m_1 - m_2}$. Thankfully, there is a key on your calculator labeled y^x (or something similar) that performs this calculation. For practice, you can confirm on your calculator that $(2.5)^5$ has a value close to 100. **(a)** How many times brighter is a star of magnitude **5** than a star of magnitude **15**? **(b)** The brightest star in the sky, Sirius, has a magnitude of **−1.5**, and the faintest star that can be seen with the unaided eye has a magnitude of about **6**. How many times brighter to the eye does Sirius appear than the faintest star?

So far we've ignored a crucial fact: stars lie at different distances from Earth. Just because one star *appears* brighter to the eye than another star does not mean that it emits more light energy; it might simply be closer than the other star. Therefore, astronomers use the lower-case magnitude symbol **m** to represent a star's *apparent brightness*, that is, the way the star appears to the eye or visually through a telescope. The uppercase magnitude symbol **M** is used to specify *absolute magnitude*, or the brightness stars would appear if they were all situated a common distance from us. Astronomers have declared this benchmark distance to be 10 parsecs from Earth, about 32.6 light-years away.

The absolute magnitude **M** is one way of expressing a star's actual light output, or *luminosity*. It effectively allows us to "line up" all stars at the same distance, so we truly can tell whether one star is a more powerful light-emitter than another. For instance, our blindingly bright Sun, if "pushed out" to the benchmark distance of 10 parsecs, would have an absolute magnitude **M** of about **4.7**; that is, it would appear rather faint to the eye. By comparison, the unremarkable-looking star 10 Lacertae, if "moved" to the benchmark 10 parsecs, would shine with an absolute magnitude **M** of **−4.8**, becoming the 3rd brightest object in the sky after the Sun and the Moon.

If a star's apparent magnitude **m** and absolute magnitude **M** are both known, then its distance can be computed by applying the inverse square law of brightness: an object's brightness diminishes with the square of the distance. Here's the procedure:

(i) Compute the difference between the star's apparent and absolute magnitudes $(m - M)$.

(ii) Convert the magnitude difference from step (i) into a brightness ratio by computing $(2.5)^{m-M}$. This result tells you how many times brighter (or fainter) the star would appear at the benchmark distance of 10 parsecs than it does now at its actual distance.

(iii) Apply the inverse square law to convert the brightness ratio from step (ii) into a distance factor. In other words, take the square root of the brightness ratio from the previous step. The result tells you how many times farther away the star is than the benchmark 10 parsecs.

(iv) Multiply 10 parsecs by the distance factor calculated in step (iii). The result is the star's actual distance in units of parsecs.

The star 10 Lacertae, noted earlier, has an apparent magnitude **m** of **4.9** and an absolute magnitude **M** that has been estimated at **−4.8**. In the following parts, you will compute its distance by the method outlined here.

3. Compute the magnitude difference **m** − **M**. (The answer is *not* 0.1.)

4. Compute the brightness ratio $(2.5)^{m-M}$.

5. Take the square root of the brightness ratio to get the distance factor.

6. Multiply the benchmark distance of 10 parsecs by this distance factor to get 10 Lacertae's actual distance in parsecs.

7. In the 1990s, the Hipparcos space telescope succeeded in measuring 10 Lacertae's parallax **p**: 0.00308 arcseconds. Recall from a previous activity that a star's distance can be computed by taking the reciprocal of its parallax, or $d = \dfrac{1}{p}$. Compute 10 Lacertae's distance from its parallax and compare it to the distance computed by the magnitude method in Part 6.

8. Before the Hipparcos space telescope, which measured the parallaxes of over a million stars, the absolute magnitude of 10 Lacertae was highly uncertain. But now that we've obtained a reliable, parallax-based distance for 10 Lacertae in Part 7, we can use that distance to compute a more accurate absolute magnitude for the star. The mathematical formula to accomplish this is: $M = m - [5 \times \log(d)] + 5$. (For the mathematical *log* function, use the key labeled *log* on your calculator.) Into the formula, substitute 10 Lacertae's apparent magnitude **m** and its "parallax distance" **d** from Part 7, and compute its new and improved absolute magnitude **M**.

9. Another common astronomical distance measure is the *light-year*, which is the distance a light beam travels through space in a year—around 6 trillion miles. **(a)** Using the conversion formula 1 parsec = 3.26 light-years, convert 10 Lacertae's parallax distance from Part 7 into light-years. **(b)** Based on the definition of the light-year, how long does it take for a light beam to travel a distance of 1 light-year through space? **(c)** In what calendar year on Earth did the light we now see from 10 Lacertae actually leave that star? **(d)** In what calendar year on Earth will the light currently leaving 10 Lacertae arrive at Earth?

How Bright Is That Star?
A Tutorial on the Magnitude System

For credit, you must show all your work.

1. (a) _____

   (b) _____

2. (a) _____

   (b) _____

3. Magnitude difference = _____

4. Brightness ratio = _____

5. Distance factor = _____

6. 10 Lacertae's distance = _____ parsecs

7. 10 Lacertae's parallax distance = _____ parsecs

8. Absolute magnitude = _____

9. **(a)** 10 Lacertae's parallax distance = _____ light-years

(b) _____ year(s)

(c) _____

(d) _____

The Realm of the Spiral Nebulae

In the early 1920s, Edwin Hubble tackled one of the most contentious questions in astronomy: are the faint spiral nebulae, which astronomers saw in vast numbers through their telescopes, complexes of gas and stars within our own Milky Way galaxy, or are they distant galaxies themselves? Hubble had already photographed the largest of the spirals, the Andromeda nebula, with the powerful 100-inch telescope on Mount Wilson in California. All he then needed to do to determine whether Andromeda was inside or outside our galaxy was to measure Andromeda's distance. But how? Even if inside the Milky Way, Andromeda lay beyond the reach of geometric and parallax distance-measuring methods. Hubble conceived an alternative path to assess Andromeda's place in the cosmos: first, he would identify a known type of star within Andromeda; next, he would check whether that star was dimmer than its like counterparts in the Milky Way; and, if it was, he would apply the inverse square law of brightness to compute the star's distance—and, hence, Andromeda's distance.

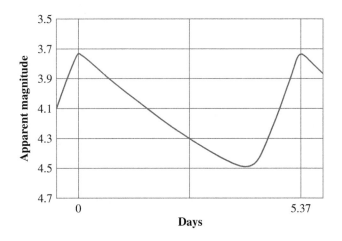

Figure 17-1
Light curve of the variable star Delta Cephei.

The key to Hubble's plan was a particular class of star called *Cepheid (seh'-fee-id) variables*. Cepheids are highly luminous stars whose outer layers expand and contract over a period of a few days to a few months. As a result, Cepheids alter their brightness in a clearly identifiable pattern. **Figure 17-1** shows the *light curve* of Delta Cephei, the first star of this type to be observed. Here, the star's brightness, or apparent magnitude, is displayed vertically and the passage of time horizontally. (Recall that the astronomical magnitude system is "backwards," so smaller magnitude numbers indicate that the star is brighter.) The period of light variation for Delta Cephei is approximately 5.37 days.

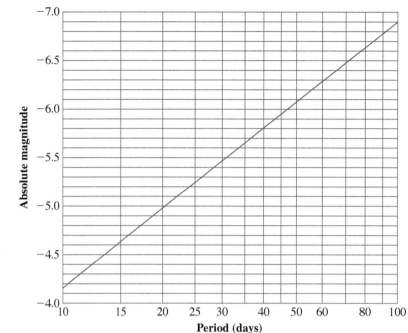

Figure 17-2
Cepheid period–
luminosity relation.

Hubble sought out Cepheids in Andromeda for two reasons: (i) Cepheids are easily identifiable by the characteristic form of their light curve; and (ii) the *period* of a Cepheid's light variation reveals the star's true light output, or *luminosity*. This *period-luminosity relation* is expressed as a graph in **Figure 17-2**. In the graph, luminosity is expressed as *absolute magnitude*, a concept introduced in the previous activity. (Absolute magnitude is the brightness a star would have if placed at the benchmark distance of 10 parsecs from Earth.)

Once Hubble determined both the Cepheids' apparent magnitude (from photographs) and absolute magnitude (from the period-luminosity graph), he applied the inverse square law of brightness to deduce the Cepheids' distance, exactly as you did in the previous activity for the star 10 Lacertae. In other words, Hubble computed how far Andromeda's Cepheids must lie for them to appear as faint as they do in photographs. In summary, to obtain the distance to the Andromeda nebula, Hubble followed these steps: (i) identified Cepheids on photographs; (ii) measured the Cepheids' apparent magnitudes and periods from these photographs; (iii) obtained the Cepheids' absolute magnitude from the graph of the period-luminosity relation; and (iv) used the inverse square law to compute the Cepheids' distance, and thus the distance of Andromeda. In this activity, you will carry out some of the steps of Hubble's groundbreaking work.

1. Hubble identified Andromeda's Cepheids by comparing their light curves to those of Cepheids in our own galaxy. Study Figure 17-1 and describe identifiable features of a Cepheid light curve that Hubble might have used.

2. The table (See **Table 17-1**) on the worksheet lists the periods and apparent magnitudes of five Cepheids Hubble discovered in the Andromeda nebula. Use the Cepheid period-luminosity relation in Figure 17-2 to obtain the absolute magnitude of each of Hubble's Cepheids. Record your answers in the appropriate column of the table. For example, according to the graph, a Cepheid with a period of 100 days would have an absolute magnitude of −6.9.

3. For each Cepheid, compute the difference between its apparent and absolute magnitudes **m** − **M** and record your results in the table.

4. Compute the corresponding brightness ratio $(2.5)^{m-M}$ for each Cepheid and record your results in the table using powers-of-ten notation.

5. Compute the "distance factor" (the square root of the brightness ratio from Part 4) and record your results in the table using powers-of-ten notation.

6. Multiply the distance factor by the benchmark distance of 10 parsecs to obtain the actual distance, in parsecs, of each Andromeda Cepheid. Record your results in the table using powers-of-ten notation.

7. **(a)** Compute the average distance of the five Cepheids by adding all five distances and dividing by 5. This is the estimated distance to the Andromeda nebula, in parsecs. **(b)** Also express Andromeda's distance in light-years by multiplying your result by 3.26. Note: Hubble's published distance was somewhat less than your answer, because he used what later proved to be a faulty period-luminosity relation.

8. Our Milky Way galaxy is a disk-shaped collection of stars about 100,000 light-years across. Looking at your result in Part 7, would Hubble have concluded that the Andromeda nebula is a star system *within* our Milky Way or a galaxy of stars lying completely *outside* the Milky Way? Explain your answer.

9. In his published paper from 1925, Hubble lists several sources of experimental uncertainty in his distance determination. What do you think these might have been? Be specific. (The phrase "human error" is not specific enough.)

The Realm of the Spiral Nebulae

For credit, you must show all your work.

1. _____
   _____
   _____

2. through 6.

Table 17-1 Data for Cepheid Variable Stars

A	B	C	D	E	F	G
Period (days)	Apparent magnitude m	Absolute magnitude M	Magnitude difference m − M	Brightness factor $(2.5)^{m-M}$	Distance factor: square root of (Column E)	Distance (parsecs) (Column F × 10)
50.17	18.4					
45.04	18.15					
41.14	18.6					
31.41	18.2					
22.03	19.0					

7. **(a)** Andromeda's distance = _____ parsecs

    (b) Andromeda's distance = _____ light-years

8. _____

    _____

    _____

9. _____

    _____

    _____

Hubble's Law—
In the Kitchen and
in the Universe

Until the 1920s, people generally believed that the universe was infinite and unchanging. However, some physicists began to realize that space itself is more than mere emtpiness; space has a definable structure and a shape. And that shape arises from the presence of matter, or equivalently, by matter's attendant force: gravity. In this revised view, a static universe is highly improbable. More likely is a universe that is either expanding or contracting. Physicists proposed that astronomers could observe the motions of galaxies and reveal the actual state of the universe. During the 1920s, there was only one telescope in the world powerful enough to conduct the necessary measurements of faint galaxies: the 100-inch reflector on Mount Wilson in California. And Edwin Hubble, who had already proven that spiral nebulae are galaxies external to our Milky Way, felt confident that he could tackle this difficult task.

To understand how Hubble intended to prove whether the universe is static or in some state of motion, let's begin with an analogy: the universe as a loaf of raisin bread. The left part of **Figure 18-1** portrays a loaf of unbaked raisin bread, showing the positions of 6 raisins, labeled **A** through **F**. It also indicates the current distance of each raisin from raisin **A**. The unit of measurement doesn't matter; let's assume it's *centimeters*, abbreviated "cm." To the right is the same loaf after it has baked and risen in the oven for 1 hour. As you can see, the loaf has expanded *uniformly* in every dimension to *twice* its original size; that is, every raisin-to-raisin distance in the original loaf is now twice as large as it was before. We'll call this increase in overall scale the *expansion factor*.

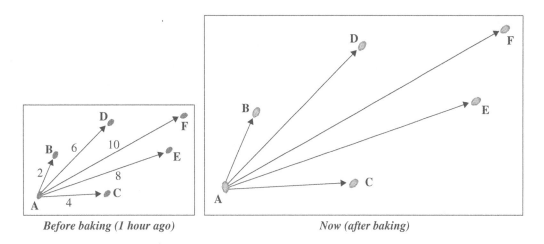

Before baking (1 hour ago)   *Now (after baking)*

Figure 18-1 Uniformly expanding raisin bread.

1. What is the distance of raisin **B** from raisin **A** now, after 1 hour of baking? (Do *not* use a ruler to measure the distance; use the number and the expansion factor given previously to formulate your measurement scale.) Record your answer on the worksheet. Then write each distance alongside its corresponding arrow in the right-hand part of Figure 18-1. Repeat for raisins **C** through **F**.

2. During the hour of baking, each raisin moved from its original position to a new position. The "velocity" of each raisin can be computed by taking the *distance the raisin moved during baking* and dividing by the *time interval* of 1 hour. Record the velocity of each raisin on the worksheet.

3. Plot your distance and velocity data from Parts 1 and 2 on the axes given on the worksheet (**Figure 18-2**).

4. Do your data points lie at least approximately along a straight line? If so, draw the straight line in the graph that best "fits" the data, that is, that has about as many data points on one side of the line as on the other. If your points do not lie approximately along a line, check your data!

5. In the baking process, the raisin bread underwent a uniform expansion. When the velocities of the raisins are plotted against their corresponding distances from raisin **A**, a straight-line relationship appears in the graph. Conversely, the straight-line relationship between the raisins' velocities and distances reveals a situation of uniform expansion. Would a straight-line relationship have appeared if you had plotted the graph based on distance and velocity measurements carried out from a different raisin than raisin **A**? Explain your answer.

6. The expansion rate of the raisin bread is represented by the steepness, or slope, of the straight line in the graph you drew in Part 3. The slope can be computed from the measurements of any two raisins, say, raisins **B** and **F**, as follows:

$$\text{slope} = \frac{(\text{velocity of raisin } \mathbf{F} - \text{velocity of raisin } \mathbf{B})}{(\text{distance of raisin } \mathbf{F} \text{ now} - \text{distance of raisin } \mathbf{B} \text{ now})}$$

Compute the slope of the line in the graph from Part 3 and record your answer on the worksheet.

Since the raisin bread is a stand-in for our universe, you can test for a uniform expansion of the universe by analogous means, in other words, by measuring and plotting the velocities and distances of galaxies, as Edwin Hubble did in the 1920s. **Table 18-1** contains the relevant data for several representative galaxies. The term "recession velocity" is used to describe galactic movement because, with few exceptions, galaxies are moving away from us, that is, they are receding.

Table 18-1 Distance and Velocity of Several Galaxies

Galaxy Name	Distance (in millions of parsecs)	Recession Velocity (in kilometers per second)
Virgo	19	1,200
Ursa Major	300	15,000
Corona Borealis	430	21,600
Bootes	770	39,300
Hydra	1,200	61,200

7. Plot the distance-velocity data for the given galaxies on the axes of **Figure 18-3** on the worksheet. Sketch the straight line that best "fits" the data points. The line does not have to pass through all the points.

8. As you should see, when the velocities of galaxies are plotted against the corresponding distances of galaxies, a straight-line relationship appears, an outcome astronomers have named Hubble's law. **(a)** Drawing a parallel to the raisin bread analogy, what does Hubble's law imply about the overall state of the universe? **(b)** Express Hubble's law in words, starting with the phrase, "The farther a galaxy is from us..." Be specific.

9. The expansion rate of the universe, called the Hubble constant, is determined by computing the slope of Hubble's law. Adapt the procedure you used in Part 6 for the raisin bread to find the value of Hubble's constant.

10. What if an "alien-Hubble" had plotted the graph based on velocity and distance measurements carried out from another galaxy? Would a straight-line, Hubble's-law relationship appear? Explain.

Hubble's Law—In the Kitchen and in the Universe

For credit, you must show all your work.

1. Distances of the various raisins from raisin A now, after one hour of baking:

   B _____ cm **C** _____ cm **D** _____ cm **E** _____ cm **F** _____ cm

2. Velocities of raisins:

   B _____ cm/hour **C** _____ cm/hour **D** _____ cm/hour **E** _____ cm/hour **F** _____ cm/hour

3.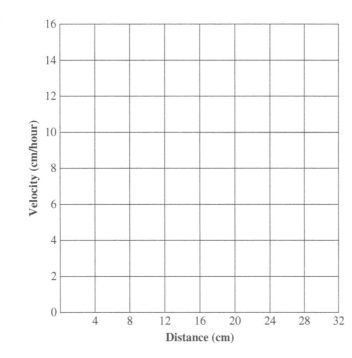

Figure 18-2
Graph of velocity vs. distance of raisins.

4. _____

5. _____

_____

_____

6. slope = _____

7.

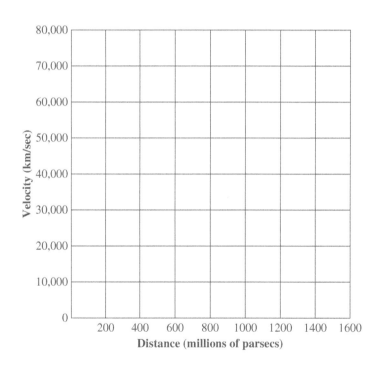

Figure 18-3
Graph of velocity vs.
distance of galaxies.

8. (a) _____

_____

(b) _____

_____

9. Hubble's constant = _____

10. _____

_____

_____

The Hertzsprung–Russell Diagram

By the early 20th century, the distances of several dozen stars had been determined reliably by the stellar parallax method. Once their distances were found, astronomers used the inverse square law of light to deduce the stars' luminosity, or absolute magnitude. Around the same time, researchers at Harvard University began to classify stars according to the appearance of their spectra, adopting letter designations that have since become standard. Soon it became evident that the major observed differences in stellar spectra stemmed, not from large differences in the chemical composition of stars, but from differences in their surface temperatures. The Harvard astronomers were then able to arrange the various spectral types in order of stellar surface temperature, from hottest to coolest: O, B, A, F, G, K, and M. Each spectral type was divided into 10 parts, indicated by a digit from 0 to 9 (for example, A0, A1, A2, ..., A8, A9, then F0, F1, F2, and so on).

Once astronomers gained a handle on a pair of stellar characteristics—luminosity and surface temperature—they wondered whether there was any correlation between the two. Working independently, astronomers Henry Norris Russell and Ejnar Hertzsprung investigated that correlation by plotting a graph of absolute magnitude versus spectral type for a sample of stars. This graph is now called the Hertzsprung–Russell diagram, or simply the HR diagram. In this activity, you will follow the path of Hertzsprung and Russell, but using modern measurements. But first, a non-astronomical introduction to the task at hand....

■ Classification, Quantification, Correlation

One way to begin a scientific analysis of a class of objects is to develop meaningful quantitative measures and classification schemes of various properties of these objects. For instance, one can graph observed data for a sample of the objects, then inspect the resulting display for correlations between various measures. To see how this might be done, let's check for possible correlations between two human characteristics, say, height and shoe size. Using the information in **Table 19-1**, fill in your own height and shoe size on the data collection sheet containing **Table 19-2** that will be passed around the class.

1. Predict what correlation(s) you might expect between people's height and their shoe size.

2. Plot the height and shoe size data collected from the class on the axes of **Figure 19-1** on the worksheet. Note: Each data point will represent one person's combination of shoe size and height.

3. Describe any correlations between height and shoe size that are evident in the graph. How do these correlations compare to what you predicted? Be specific.

■ Plotting the HR Diagram

Table 19-3 lists physical data for two categories of stars: the brightest stars in the sky; and the nearest stars, those within about 10 light-years of our solar system.

4. For each of the data sets in **Table 19-3**, plot the absolute magnitude versus the spectral type on the axes of **Figure 19-2** on the worksheet. Use a "•" symbol to represent the brightest stars and an "**x**" to represent the nearest stars. Note: A letter "D" before the regular spectral-type letter indicates a collapsed star known as a white dwarf. And remember, the *smaller* magnitude number, the *brighter* the star.

5. Observe that the majority of both the nearest and the brightest stars fall along a diagonal band in the HR diagram. Astronomers have named this band the *main sequence*. Draw a line in the graph that traces the main sequence. What are the general luminosity and temperature properties of stars **(a)** in the upper part of the main sequence? **(b)** in the lower part of the main sequence?

6. Another group of stars falls in the upper right portion of the HR diagram. These are *red giant* and *red supergiant* stars. Draw a circle around these stars. **(a)** What are the general luminosity and temperature properties of giants and supergiants? **(b)** Can you explain why astronomers believe these stars must be very large? (Hint: A cool star radiates less energy per square foot of surface area than does a hot star.)

7. In the lower left portion of the HR diagram are white dwarf stars. Circle these. **(a)** What are the general luminosity and temperature properties of white dwarfs? **(b)** Can you explain why astronomers believe these stars must be very small? (Hint: A hot star radiates more energy per square foot of surface than does a cool star.)

8. **(a)** How do the overall distributions of the brightest and the nearest stars differ from one another in the HR diagram? **(b)** What does this imply about the fundamental properties of each of these two categories of stars?

9. Which star *appears* brightest in the night sky?

10. **(a)** Which star is *farthest* from the solar system? **(b)** Why does such a faraway star nevertheless appear so bright in the night sky?

Table 19-1   Shoe size and height data guidelines

If your *shoe size* is …		Then write the number below in the "shoe size" column	If your *height* is… (round up half-inches)	Then write the number below in the height column
Men	Women			
3 1/2	5	5	Less than 5 ft 0 in.	59
4	5 1/2	5.5	5 ft 0 in.	60
4 1/2	6	6	5 ft 1 in.	61
5	6 1/2	6.5	5 ft 2 in.	62
5 1/2	7	7	5 ft 3 in.	63
6	7 1/2	7.5	5 ft 4 in.	64
6 1/2	8	8	5 ft 5 in.	65
7	8 1/2	8.5	5 ft 6 in.	66
7 1/2	9	9	5 ft 7 in.	67
8	9 1/2	9.5	5 ft 8 in.	68
8 1/2	10	10	5 ft 9 in.	69
9	10 1/2	10.5	5 ft 10 in.	70
9 1/2	11	11	5 ft 11 in.	71
10	11 1/2	11.5	6 ft 0 in.	72
10 1/2	12	12	6 ft 1 in.	73
11	12 1/2	12.5	6 ft 2 in.	74
11 1/2	13	13	6 ft 3 in.	75
12	13 1/2	13.5	6 ft 4 in.	76
12 1/2	14	14	More than 6 ft 4 in.	77
13	14 1/2	14.5		
13 1/2	15	15		

Table 19-2   Shoe size and height data collection sheet

Fill in your shoe size and height using the information in **Table 19-1**

	Shoe Size	Height	Shoe Size	Height
example →	13	71		

Brightest Stars (•)

Name	Distance (light-years)	Apparent magnitude m	Absolute magnitude M	Spectral type
Sun	–	–26.7	4.8	G2 V
Sirius A	8.6	–1.46	1.4	A1 V
Canopus	74	–0.72	–2.5	A9 II
Alpha Centauri A	4.3	–0.01	4.4	G2 V
Alpha Centauri B	4.3	1.33	5.7	K0 V
Arcturus	34	–0.04	0.2	K2 III
Vega	25	0.03	0.6	A0 V
Capella	41	0.08	0.4	G6 III
Rigel	~1400	0.12	–8.1	B8 Ia
Procyon A	11.4	0.38	2.6	F5 IV–V
Achernar	69	0.46	–1.3	B3 V
Betelgeuse	~1400	0.50	–7.2	M2 Iab
Hadar	320	0.61	–4.4	B1 III
Acrux	510	0.76	–4.6	B1 I
Altair	16	0.77	2.3	A7 V
Aldebaran	60	0.85	–0.3	K5 III
Antares	~520	0.96	–5.2	M2 Iab
Spica	220	0.98	–3.2	B1 V
Pollux	40	1.14	0.7	K0 III
Fomalhaut	22	1.16	2.0	A3 V
Becrux	460	1.25	–4.7	B1 III
Deneb	1500	1.25	–7.2	A2 Ia
Regulus	69	1.35	–0.3	B7 V
Adhara	570	1.50	–4.8	B2 II
Castor	49	1.57	0.5	A1 V
Gacrux	120	1.63	–1.2	M4 III

Nearest Stars (x)

Name	Distance (light-years)	Apparent magnitude m	Absolute magnitude M	Spectral type
Sun	–	–26.7	4.8	G2 V
Proxima Centauri	4.2	11.05	15.5	M6 V
Alpha Centauri A	4.3	–0.01	4.4	G2 V
Alpha Centauri B	4.3	1.33	5.7	K0 V
Barnard's Star	6.0	9.54	13.2	M4 V
Wolf 359	7.7	13.53	16.7	M6 V
Lalande 21185	8.2	7.50	10.5	M2 V
UV Ceti A	8.4	12.52	15.5	M6 V
UV Ceti B	8.4	13.02	16.0	M6 V
Sirius A	8.6	–1.46	1.4	A1 V
Sirius B	8.6	8.3	11.2	D(A2)
Ross 154	9.4	10.45	13.1	M4 V
Ross 248	10.4	12.29	14.8	M5 V
Epsilon Eridani	10.8	3.73	6.1	K2 V
Lacaille 9352	10.8	7.34	9.8	M2 V
Ross 128	10.9	11.10	13.5	M4 V
61 Cygni A	11.1	5.2	7.6	K4 V
61 Cygni B	11.1	6.03	8.4	K5 V
Epsilon Indi	11.2	4.68	7.0	K3 V
BD +43 44	11.2	8.08	10.4	M1 V
BD +43 44	11.2	11.06	13.4	M4 V
Procyon A	11.4	0.38	2.6	F5 IV–V
Procyon B	11.4	10.7	13.0	D(A)
BD +59 1915 A	11.6	8.90	11.2	M3 V
BD +59 1915 B	11.6	9.69	11.9	M4 V
CoD –36 15693	11.7	7.35	9.6	M1 V

Table 19-3 Data for the brightest and the nearest stars

The Hertzsprung–Russell Diagram

For credit, you must show all your work.

1. _____
   _____

2.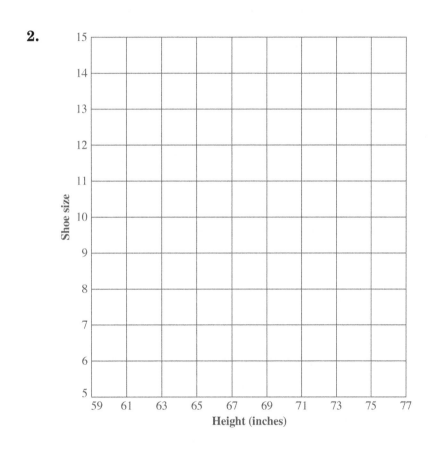

Figure 19-1 Graph of shoe size vs. height.

3. _____

_____

4.

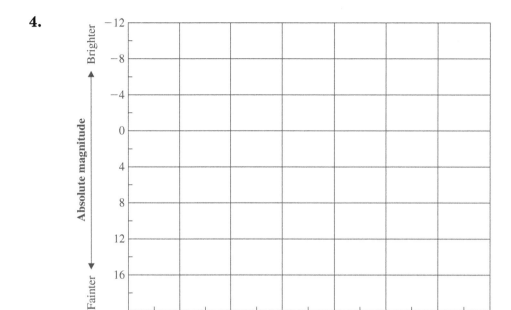

Figure 19-2 Graph of absolute magnitude vs. spectral type.

5. (a) _____

(b) _____

6. (a) _____

(b) _____

_____

_____

7. (a) _____

(b) _____

_____

_____

8. (a) _____

_____

_____

(b) _____

_____

_____

9. _____

10. (a) _____

(b) _____

_____

Binary Stars and Stellar Masses

"The visibility of countless stars is no argument against the invisibility of countless others."

>—Astronomer Friedrich Bessel, in an 1846 letter, after concluding that the star Sirius is paired with an unseen companion star.

"Prof. Bond [at Harvard] communicates the discovery of a Companion of Sirius, made on the evening of Jan. 31 by Mr. Clark, with his new object-glass of 18½ inches aperture."

>—*Monthly Notices of the Royal Astronomical Society*, March 14, 1862.

In 1803, the eminent English astronomer William Herschel concluded that at least some double stars—pairs of stars close together in the sky—are, in fact, binary systems: stars in orbit around one another. Astronomers realized that observations of binary stars would allow them to determine a stellar property that had been otherwise measurable only for the Sun: a star's mass. The basis of stellar mass measurement dates to the mid-1600s, when Isaac Newton derived Kepler's laws from his own Universal Law of Gravitation. The upshot is that any orbiting system of objects bound together by gravity will conform to Kepler's laws, whether that system is the Moon orbiting Earth, a planet orbiting the Sun, or a binary star whose members orbit each other. Therefore, careful measurements of the orbital radius and the orbital period of a binary star could lead to a determination of stellar mass via Kepler's Third Law, which relates these three quantities.

Some four decades after Herschel's work, German astronomer Friedrich Bessel found that the bright star Sirius displays a wobbling movement in the sky over many years, as though it were dragging along a massive, but invisible, companion star. That furtive body, all but lost in Sirius's glare, was discovered accidentally in 1862 by Cambridge, Massachusetts, telescope maker Alvan Graham Clark while testing a new telescope. It was a star of exquisite strangeness, unlike any that had ever been seen. It was massive enough to tug heavy Sirius from its regular path, yet barely luminous enough to be seen through a powerful telescope—a *white dwarf* star.

In this activity, you will "observe" the Sirius star system and use Kepler's Third Law to determine the mass of Sirius itself, as well as that of its elusive companion, known as Sirius B. Kepler's Third Law is typically written in the familiar short form $P^2 = a^3$. As noted in a previous activity, its more complete form is $M \times P^2 = a^3$ or, with a slight algebraic manipulation, $M = a^3 / P^2$. Here **M** stands for the combined mass of the orbiting stars (Sirius plus its companion) expressed as a multiple of the Sun's mass (a unit called a "solar mass"). The period **P** is expressed in years and the orbital radius **a** in astronomical units (AUs). In the steps below, you will determine first the orbital period, then the orbital radius, and finally the combined and individual masses of the stars.

■ Orbital Period

Figure 20-1 shows the orbital path of Sirius B around Sirius itself, which is represented by a large dot inside the orbit. (In reality, the two stars orbit each other, but for clarity the orbit is displayed as though Sirius is fixed in place and Sirius B moves around it.) This orbit is based on decades of observation, where succeeding positions of Sirius B are indicated by a series of dots, each labeled with the year of observation.

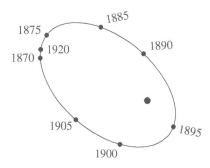

Figure 20-1
Orbital path of Sirius B.

1. (a) Study the labeled years of observation and use them to estimate the orbital period of Sirius B, in years. **(b)** Explain how you arrived at your answer.

■ Orbital Radius

The orbital radius of the Sirius system, as seen from Earth, spans an angle of about 7.6 arcseconds. (Remember, an arcsecond is $\frac{1}{3600}$ of a degree.) However, Kepler's Third Law requires, not the angular span, but the *actual* orbital radius **a**, in AUs. The sector equation that you used in previous activities can be used here as well to make the needed conversion from angle to radius. In the sector equation, $a = \dfrac{(d\theta)}{57.3}$, θ represents the observed angular span of the Sirius system's orbital radius, in degrees; **a** represents the actual orbital radius, in AUs; and **d** represents Sirius's distance, also in AUs. (See **Figure 20-2**.) The steps that follow will guide you through the procedure.

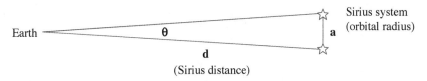

Figure 20-2 Viewing the Sirius binary system from Earth.

2. As stars go, Sirius is relatively close by. Therefore it's no surprise that its parallax was measured as early as the 1840s and is now known quite accurately: 0.379 arcseconds. Use the parallax equation from a previous activity to compute the distance of Sirius: $d = \dfrac{1}{p}$, where **d** is the distance, in parsecs, and **p** is the parallax, in arcseconds. (Don't confuse the period symbol **P** with the parallax symbol **p**.)

3. Convert your distance of Sirius from parsecs to AUs. Use the fact that there are 206,265 AUs in one parsec.

4. The angular span **θ** of the Sirius system's orbital radius is 7.6 arcseconds. Convert **θ** from arcseconds to degrees. Use the fact that there are 3600 arcseconds in one degree.

5. Use the sector equation cited previously in this activity to compute the actual orbital radius **a** of the Sirius system, in AUs.

■ Combined Mass of the Sirius System

6. Substitute your computed values for period **P** and orbital radius **a** into Kepler's Third Law $M = \dfrac{a^3}{P^2}$ and solve for the combined mass **M** of Sirius and its companion, in solar masses.

■ Masses of the Individual Stars

Using some elementary physics, it's possible to take the combined mass of the Sirius system and separate out the individual masses of Sirius and its companion, Sirius B. **Figure 20-3** shows the gradual movement of the Sirius system through space over many decades. The large dots indicate Sirius's position and the smaller dots the position of Sirius B. Corresponding dots for each year of observation are connected by a line segment. Tracing the figure from top to bottom, notice that as the orbiting system moves through space, the two stars swing around an imaginary straight line between them. This line defines the system's *center of mass*. Here's where the physics comes in. The heavier star (Sirius) always lies closer to the center-of-mass line than the lighter star (Sirius B). And the ratio of Sirius's mass to that of its companion, $\dfrac{M_s}{M_c}$, will be equal to the ratio of the companion's distance from the center-of-mass line to that of Sirius, $\dfrac{r_c}{r_s}$. That is: $\dfrac{M_s}{M_c} = \dfrac{r_c}{r_s}$. A similar principle explains why, in order to balance a seesaw, the heavier person must sit closer to the central pivot than the lighter person.

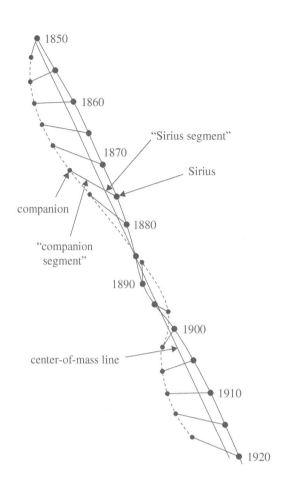

Figure 20-3
The paths of Sirius and its
companion through space.

7. To begin to solve the equation, first we have to measure r_c and r_s. In Figure 20-3, choose any one of the line segments that connect the stars for a given year of observation. Align a slip of paper alongside the segment. With your pencil, mark off the length of only that portion of the segment that extends from the center-of-mass line to Sirius. This is the "Sirius segment" r_s. Now slide your marked paper over to the portion of the segment that connects the center-of-mass line to the companion star, Sirius B. Approximately how many of these marked "Sirius segments" does it take to trace out the corresponding "companion segment?" The answer to this question is precisely the ratio we're looking for, $\dfrac{r_c}{r_s}$ and, according to the above equation, is also equal to the mass ratio $\dfrac{M_s}{M_c}$. You should carry out this procedure for several line segments in Figure 20-3 to confirm that your answers are consistent.

8. From Part 6, we know the sum of the masses of Sirius and its companion. And from Part 7, we know the ratio of the masses of the individual stars. Now, using simple algebra or trial and error, you can compute the individual masses of Sirius and Sirius B, in solar masses. For example, suppose $a + b = 12$ and $a/b = 3$. What are a and b? Solution: $a/b = 3$ becomes $a = 3b$; substitute into the first equation $a + b = 3b + b = 4b = 12$, so $4b = 3$, and therefore $a = 9$.

Binary Stars and Stellar Masses

For credit, you must show all your work.

1. (a) orbital period = _____ years

(b) _____
_____

2.

distance of Sirius **d** = _____ parsecs

3.

distance of Sirius **d** = _____ AUs

4.

angular span **θ** = _____ degrees

5.

orbital radius **a** = _____ AUs

6.

combined mass **M** = _____ solar masses

7.

companion segment = _____ Sirius segments

8.

mass of Sirius = _____ solar masses

mass of companion = _____ solar masses

Appendix: Mathbits

■ A Taste of Trigonometry

There are many applications in astronomy that involve right triangles, which have one angle that measures 90 degrees. Trigonometry describes the mathematical relationship between the measures of the angles and the lengths of the sides of a right triangle. It does this by introducing a number of trigonometric functions.

The basic trigonometric ("trig") functions of a right triangle are *sine*, *cosine*, and *tangent*, abbreviated **sin**, **cos**, and **tan**. For example, the sine of 30 degrees is written **sin** 30; the cosine of the angle θ ("theta") is written **cos** θ. Each trig function is defined in terms of the lengths of the triangle's sides: the side opposite the angle; the side adjacent to the angle; and the hypotenuse, which is the side opposite the 90-degree angle and also the longest side.

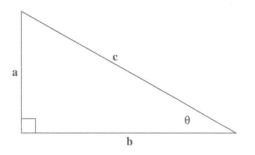

Figure App-1
A right triangle.

Using the labels in **Figure App-1**, here are the definitions of the three basic trig functions:

$$\sin\theta = \frac{\text{opposite side}}{\text{hypotenuse}} = \frac{a}{c}$$

$$\cos\theta = \frac{\text{adjacent side}}{\text{hypotenuse}} = \frac{b}{c}$$

$$\tan\theta = \frac{\text{opposite side}}{\text{adjancent side}} = \frac{a}{b}$$

For example, if side **a** is 3 inches long, and side **b** is 4 inches long, and the hypotenuse, side **c**, is 5 inches long, then $\sin\theta = 3/5 = 0.6$, $\cos\theta = 4/5 = 0.8$, and $\tan\theta = 3/4 = 0.75$. For such a triangle, the expression $5\sin\theta = (5) \times (0.6) = 3$; and $10\tan\theta = (10) \times (0.75) = 7.5$. Alternatively, if you are instead given the measure of the angle θ rather than the lengths of the sides, you can simply use your calculator to compute the numerical value of a trig function. There are keys on your calculator to do this, labeled (duh!) **sin**, **cos**, and **tan**.

Sometimes you are given, say, the sine of an angle and you have to figure out the measure of the angle itself. For example, what is the angle θ whose sine has the value 0.6? In a sense, you are trying to "undo" the trig function to obtain the angle. For this purpose, your calculator has keys labeled $\sin^{-1}$, $\cos^{-1}$, and $\tan^{-1}$, which are read *inverse sine, inverse cosine,* and *inverse tangent.* On some calculators, you may have to press a special *2nd function* key to activate the inverse trig keys. For practice, use your calculator to find the angle whose (a) tangent equals 1; (b) whose cosine equals 0.3.

A final word: for all the activities in this book, your calculator must be set up to accept values of angles in units of degrees, not in units of radians. Ask someone for help if you don't know how to set up your calculator.

■ Scientific Notation

Given the enormous scale of sizes, masses, distances, densities, and the sheer count of objects in the universe, astronomers routinely deal with very large numbers. Rather than writing out long strings of digits, scientists have developed a shorthand method to express large numbers—scientific notation, also known as powers-of-ten notation. In this system, every number consists of a two-part mathematical expression: a coefficient between 1 and 10; and a multiplier that is a power of 10, that is, 10 raised to an integer exponent. The powers of ten are:

$$10 = 10^1$$
$$100 = 10^2$$
$$1000 = 10^3$$
$$10{,}000 = 10^4$$
$$100{,}000 = 10^5$$
$$1{,}000{,}000 = 10^6\text{, and so on.}$$

In scientific notation, the number 510,000 is represented as 5.1×10^5; the exponent 5 is the number of places the decimal point in 5.1 has to be moved to the right to restore the original number 510,000. Similarly, the radius of Earth's orbit—93,000,000 miles—is written as 9.3×10^7 miles. And Earth's mass—about 5,973,600,000,000,000,000,000,000 kg—abbreviates to 5.9736×10^{24} kg in scientific notation.

By the way, on many calculator displays, the exponent portion of a powers-of-ten number is indicated by the letter "E." Thus, 6.2×10^{20} might appear on your calculator as 6.2 E20. To input a powers-of-ten number into your calculator, key in the coefficient portion of the number, then press the exponent key (labeled something like "EE" or "EXP"), followed by the exponent.

CPSIA information can be obtained
at www.ICGtesting.com
Printed in the USA
BVHW012149070922
646546BV00012B/285